Lecture Notes in Economics and Mathematical Systems

Founding Editors:

M. Beckmann
H. P. Künzi

Managing Editors:

Prof. Dr. G. Fandel
Fachbereich Wirtschaftswissenschaften
Fernuniversität Hagen
Feithstr. 140/AVZ II, 58084 Hagen, Germany

Prof. Dr. W. Trockel
Institut für Mathematische Wirtschaftsforschung (IMW)
Universität Bielefeld
Universitätsstr. 25, 33615 Bielefeld, Germany

Co-Editors:

C. D. Aliprantis

Editorial Board:

A. Basile, A. Drexl, G. Feichtinger, W. Güth, K. Inderfurth, P. Korhonen,
W. Kürsten, U. Schittko, P. Schönfeld, R. Selten, R. Steuer, F. Vega-Redondo

Springer
*Berlin
Heidelberg
New York
Hong Kong
London
Milan
Paris
Tokyo*

Cornelia Neff

Corporate Finance, Innovation, and Strategic Competition

 Springer

Author

Dr. Cornelia Neff
ING BHF-BANK AG
Bockenheimer Landstr. 10
60323 Frankfurt/Main
Germany

Cataloging-in-Publication Data applied for
Bibliographic information published by Die Deutsche Bibliothek
Die Deutsche Bibliothek lists this publication in the Deutsche Nationalbibliografie;
detailed bibliographic data is available in the Internet at http://dnb.ddb.de

ISSN 0075-8450
ISBN 3-540-44294-4 Springer-Verlag Berlin Heidelberg New York

This work is subject to copyright. All rights are reserved, whether the whole or part
of the material is concerned, specifically the rights of translation, reprinting, re-use
of illustrations, recitation, broadcasting, reproduction on microfilms or in any other
way, and storage in data banks. Duplication of this publication or parts thereof is
permitted only under the provisions of the German Copyright Law of September 9,
1965, in its current version, and permission for use must always be obtained from
Springer-Verlag. Violations are liable for prosecution under the German Copyright
Law.

Springer-Verlag Berlin Heidelberg New York
a member of BertelsmannSpringer Science+Business Media GmbH

http://www.springer.de

© Springer-Verlag Berlin Heidelberg 2003
Printed in Germany

The use of general descriptive names, registered names, trademarks, etc. in this
publication does not imply, even in the absence of a specific statement, that such
names are exempt from the relevant protective laws and regulations and therefore
free for general use.

Typesetting: Camera ready by author
Cover design: *Erich Kirchner*, Heidelberg

Printed on acid-free paper   SPIN: 10893968   55/3142/du   5 4 3 2 1 0

Preface

Industrial organization considers strategic behavior of firms in the product market. Firms compete in prices and invest in innovation activities in order to gain market shares. In this book I will investigate the financial decisions of innovative firms if financial markets are imperfect due to asymmetric information. I will demonstrate how financial market imperfections interact with strategic competition of firms in the product market. The tool to analyze these strategic interactions is non-cooperative game theory.

This book was written while I was assistant professor at the department of economic theory at the University of Tübingen. I wish to thank my supervisor Manfred Stadler for having suggested this interesting research subject. He introduced me to the exciting field of industrial organization and the economics of information. As my supervisor, he gave me great freedom to pursue my research.

I would like to thank my colleagues Jürgen Volkert, Andreas Scheuerle, Stephan Hornig, Rüdiger Wapler and Leslie Neubecker for the nice and friendly atmosphere at our office. Katharina Wichert, Frank Breitling, Andrea Schrage, Stephan Göbel, Christina Schumacher, Alexandra Zaby, Tina Bach-Adetunji and Vanessa Steinmeier provided valuable research assistance.

I am grateful to Werner Neus for being my second supervisor. He showed me that the department of banking and business administration and the department of economic theory share common views on financial markets under asymmetric information.

The present book summarizes the research that I undertook for the research project "Financial Markets and Innovation Activities" of the German Science Foundation (DFG). This research project was part of the long-term DFG research program (1996–2001) on "Industrial Organization and Input Markets". I am grateful to the German Science Foundation for financial support.

In addition to this, I wish to thank Konrad Stahl and Hans-Jürgen Ramser for the superb organization of the semi-annual conferences of our DFG research program in Heidelberg. Here, we discussed the theoretical and empirical results of our research. Moreover, I would like to express my thanks to the DFG, the Center for Economic Policy Research, London, and the Zentrum für Europäische Wirtschaftsforschung, Mannheim, for the bi-annual joint conferences held in Mannheim. These conferences served as international forum to present our work. I am also thankful to the discussants at the annual meetings of the European Economic Association 1998 in Santiago de Compostella and 1999 in Bozen, and to the participants of the doctoral colloquium of the Studienstiftung des Deutschen Volkes in 1998.

I would like to thank Urs Schweizer for organizing the International Summer School of the European Doctoral Program 1998 and 1999 on the castle of Schönburg at Oberwesel. These courses have greatly influenced my understanding of contract theory and applied game theory.

For advice, encouragement, and criticism on specific chapters I am particularly indebted to Horst Albach, Ulrich Binkert, Urs Birchler, Monika Bütler, Annette Boom, Hans Hirth, Stephan Hornig, Franz Hubert, Annette Kirstein, Werner Neus, Rolf Niedermeier, Thomas Pfeiffer, Martin Ruckes, Vera Schubert, Manfred Stadler, and Uwe Walz.

As a researcher in economic theory, I have benefited greatly from conversations with practitioners, i.e. bankers, start-up entrepreneurs and venture capitalists. On behalf of ING BHF-BANK, Frankfurt, I wish to thank Jürgen Kanne and Manfred Moll for the very fruitful and open discussions on corporate finance and bank-firm relationships. Nick Money-Kyrle, Clemens Busch, and Gabriele Egger of BHF Private Equity, Frankfurt, provided excellent advice on private equity financing. Holger Dietrich and his colleagues from Blue Elephant Systems, Herrenberg, gave me valuable insights into the adventure of founding a company, drafting a business plan, and raising venture capital.

The University of Tübingen provided a highly intellectual and truly inspiring atmosphere for the realization of this book. Ulrich Binkert, my sister Claudia, and my grand-parents were a source of steady encouragement. I also would like to thank my friends and climbing partners Antonella Raddi, Brigitte Pflug, Guido Sobbe, and Holger Dietrich for the great climbs that we did together on the weekends. Last but most, I am indebted to my mother. She financially supported my university studies which laid the foundations of this book.

Frankfurt am Main, September 2002   Cornelia Neff

Table of contents

1 Introduction 1

2 Financial structure and strategic competition 4
    2.1 Financial structure decision of an individual firm 5
        2.1.1 Modigliani Miller proposition 5
        2.1.2 Theories of optimal capital structure 6
    2.2 Financial structure and product market competition 7
        2.2.1 Optimal capital structure in a static framework 8
            2.2.1.1 Limited liability effect and price competition 8
            2.2.1.2 Limited liability effect and price and capacity competition 14
        2.2.2 Capital structure in a dynamic framework 22
            2.2.2.1 Price competition and long-term debt 22
            2.2.2.2 Impact of consumer switching costs 27
            2.2.2.3 Product quality and long-term debt 29
            2.2.2.4 Predatory pricing 34
            2.2.2.5 Summary 35
        2.2.3 Capital structure and market entry 36
            2.2.3.1 Assumptions 36
            2.2.3.2 Output stage 39
            2.2.3.3 Financial stage 41
            2.2.3.4 Discussion 41
    2.3 Empirical findings 42
    2.4 Conclusion 43

3 Credit financing and strategic competition 46
    3.1 Introduction: Contract theory 47
        3.1.1 General ideas and definitions 47
        3.1.2 The revelation principle 48
    3.2 Credit financing: The individual firm-bank relationship 49
        3.2.1 Moral hazard 50
            3.2.1.1 Costly state verification and standard debt contract ... 50
            3.2.1.2 Renegotiation of the standard debt contract 54

		3.2.2	Multi-period credit contracting		56
			3.2.2.1	Dynamic moral hazard	56
			3.2.2.2	Long-term contracting when costly state verification is not possible	57
		3.2.3	Discussion		61
	3.3	Credit financing and product market competition			62
		3.3.1	Credit financing and price competition		63
			3.3.1.1	Credit financing, price competition, and consumer switching cost	63
			3.3.1.2	Credit financing, market power of banks, and price competition	69
		3.3.2	Credit financing, innovation, and product market competition		76
			3.3.2.1	Introduction	76
			3.3.2.2	A dynamic game of innovation and price competition	77
			3.3.2.3	Optimal debt contract for one externally financed firm	87
			3.3.2.4	Optimal contracts when both firms need external debt financing	94
			3.3.2.5	Predation	100
		3.3.3	Discussion		101
	3.4	Conclusion			103
	3.5	Appendix			104

4 Venture capital financing and strategic competition ... 106

	4.1	Introduction			107
		4.1.1	The financial growth cycle		107
		4.1.2	Characteristics of venture capital financing		110
	4.2	Venture capital financing – the individual firm's perspective			113
		4.2.1	Allocation of control rights		114
			4.2.1.1	Assignment of control rights	114
			4.2.1.2	The search for a professional management	122
			4.2.1.3	Efficient intervention by the venture capitalist	123
			4.2.1.4	Discussion	125
		4.2.2	Allocation of ownership rights		125
			4.2.2.1	Moral hazard and financial instruments	126
			4.2.2.2	Double moral hazard	129
			4.2.2.3	Contracting with a venture capitalist and an outside investor	135
			4.2.2.4	Dynamic contracting, moral hazard, and learning	141
			4.2.2.5	Discussion	150

4.3	Venture capital financing and product market competition			151
	4.3.1	Assumptions		153
		4.3.1.1	The basic model	157
		4.3.1.2	Innovation, entry, and competition in a two-period framework	158
	4.3.2	Venture capital contracting		161
		4.3.2.1	Short-term contracting	162
		4.3.2.2	Long-term contracting	168
		4.3.2.3	Discussion of the results	176
	4.3.3	Strategic reactions of the incumbent		178
		4.3.3.1	Innovation and cost reduction	178
		4.3.3.2	Predation	184
	4.3.4	Discussion		189
4.4	Conclusion			191
4.5	Appendix			192

5 Conclusion 200

References 207

Figures 214
Tables 215
List of symbols 216

1 Introduction

Corporate finance models and the theory of industrial organization have long been two separate fields of research. An integrating analysis is missing so far, although it is well known that financial market imperfections spill over to the product market and influence, in turn, the competitive strategies of firms.

The aim of the present book is to bridge this gap. We integrate the financing decisions of firms into industrial organization models of product market competition.

On the product market side, we focus on dynamic price competition. In addition to this, firms engage in R&D activities. Innovation can be in the form of product or process innovation. In case of process innovation, successful firms reduce their marginal production cost. In case of product innovation, a firm enters a market with a new product variant.

We assume that financial markets are imperfect due to asymmetric information. If a firm needs external financing, it signs a financial contract with an investor. The contract must be designed in a way that the information problem is circumvented. We consider debt and equity contracts. Debt is provided in the form of bank loans, while equity is provided in the form of venture capital.

Our main research subject is to investigate the consequences of these financial debt and equity contracts on the strategic competition in the product market.

We proceed as follows: In each chapter, we first describe the corporate finance decision of an individual firm. Secondly, we analyze strategic competition between firms on the industry level and investigate the impact of the financial decision on the firms' innovation and pricing strategies. As it is standard in the theory of industrial organization, we also consider market entry and exit. The framework of our analysis is dynamic.

Chapter 2 investigates the capital structure decision of firms. The chapter serves as preparation for the subsequent chapters 3 and 4, where we explicitly analyze equity and debt contracts of firms. In the first part of chapter 2, we briefly describe the various issues of optimal capital structures of firms, which evolved as a reaction to the Modigliani and Miller proposition.

In the second part, we consider the impact of a firm's capital structure decision on product market competition. The key assumption here is that firms which take up debt for strategic reasons are protected by limited liability. This limited liability effect alters the firms' competitive strategies. On the product market, we focus on price competition and analyze the firms' behavior in a static and dynamic framework. Moreover, we include capacity considerations, product quality decisions, and predation strategies into our analysis. Finally, we turn to the capital structure choice of a young firm that attempts to enter a market, where it faces price competition with an incumbent firm.

The third part of chapter 2 is dedicated to empirical findings. We present several empirical studies which investigate the pricing strategies of firms that suddenly change their capital structure by increasing their debt capacities. These empirical results from leveraged-buy-out activities help us to evaluate the theoretical models on capital structure decision and product market competition presented above.

In the subsequent chapters 3 and 4, we leave the simple debt versus equity choice aside. Instead, we analyze the explicit financial contracts between the firm and its debt or equity investor.

In the introduction to chapter 3, we briefly describe the main ideas of contract theory. The remainder of chapter 3, then, focuses on bank loan financing of two well-established firms. In section (3.2), we analyze the individual firm-bank relationship and review the standard models of ex-post asymmetric information in the credit market. Then, we show that long-term debt contracts mitigate the moral hazard problem, because the bank can deny follow-up financing for a firm whose reported profits are too low.

Our analysis on the industry level (3.3) combines long-term debt contracts with product market competition. We analyze two models of price competition and derive the optimal debt contracts. Then, we present our own work on dynamic price competition and cost-reducing innovation. We investigate the impact of debt contracts on strategic competition and cost-reducing innovation, if either one firm or both need external financing. Moreover, we show how the debt contracts influence the R&D intensities within the industry. Finally, we show how the financial contract induces a self-financed firm to prey upon its debt-financed rival.

Chapter 4 investigates venture capital financing – as a form of equity financing - and product market competition. We assume that a wealth-constrained entrepreneur owns an innovative product idea and seeks external funding. Since bank loan financing is typically denied for a young firm, the entrepreneur has to find an alternative form of funding. In section (4.1) we describe the financing alternatives for firms as they move along their financial growth cycle. We show that venture capital companies provide funds for high-risk innovation projects. In exchange for the provision of capital, they obtain a share of the project's returns.

Venture capital financing has become a booming area of research. However, only the bilateral relationship between the venture capital company and the start-up entrepreneur has been analyzed so far. In the second part of chapter 4, we review the literature on how control and ownership rights are distributed in venture capital contracting. The financial relationship is characterized by asymmetric information between the entrepreneur and the venture capital investor.

None of these papers, though, has investigated the competitive environment in which the young firm operates. Our analysis on the industry level is the first to combine venture capital financing with product market competition. In section (4.3) we explicitly formalize the market entry of a young, venture capital backed firm and its subsequent price competition with an incumbent. We derive the optimal financial contract and show how it depends on the profit opportunities of the innovation and competition game. Moreover, we derive how the financial contracting influences the competitive structure of the industry.

In the last part of chapter 4, we consider the possibility that the incumbent engages in predatory activities in order to deter the young firm's market entry. We analyze the impact of the incumbent's strategy on the young firm's optimal financial contract. The theoretical findings are supported by a short case study on strategic competition between self-financed incumbents and a venture capital backed start-up firm.

The conclusion reviews our contribution to research. We compare the features of our optimal debt and equity contracts, pointing out their similarities and differences. Then, we summarize how these financial contracts influence the firms' competitive strategies and innovation activities in the product market.

2 Financial structure and strategic competition

"Why financial structure matters"

Joseph Stiglitz (1988)

In this chapter we point out that in contrast to the Modigliani-Miller Proposition, financial structure decisions of firms do matter. To show this, we first consider the financial structure decision of an individual firm (2.1). We summarize the Modigliani and Miller result that is derived when financial markets are perfect. Then, we briefly describe various issues of the theory of optimal capital structure.

In the section (2.2), we examine the impact of capital structure decisions on product market competition between two incumbent firms. The financial decision of each firm consists here of a simple debt versus equity choice. We see that the limited liability effect alters the product market strategies if a firm takes up debt. In case of price competition, taking up debt softens the intensity of competition between firms. In case of price and capacity competition, by contrast, firms will switch to a more aggressive output market strategy if they take up debt.

In the second part of section (2.2), we investigate the impact of capital structure decisions on dynamic price competition. Firms anticipate subsequent rounds of competition and take intertemporal aspects – like investments in the acquisition of market shares – into account. This gives us a richer and more realistic picture of product market competition.

In the last part of section (2.2) we investigate market entry. We analyze the capital structure choice of a firm that attempts to enter a market, where it faces price and capacity competition with an incumbent firm. We show that upon observing the young firm's capital structure, the incumbent has incentives to prey and to alter its product market strategy.

Section (2.3) is dedicated to empirical findings about the firm's capital structure decisions and their impact on product market competition. Most studies investigate pricing strategies after one or several firms in that industry have undergone a leveraged buyout. Leveraged buyout transactions dramatically increase the debt-equity ratio of a firm. These empirical findings help us to evaluate the theoretical models of the previous sections.

2.1 Financial structure decision of an individual firm

In this section we consider the corporate finance decision of an individual firm, which owns a risky investment project. The choice of financial instruments is related to the total market value of the firm. We start with the famous Modigliani-Miller proposition and briefly describe the literature which evolved on the *non-irrelevance* of capital structure thereafter. Then, we present more in detail the financial structure decision of an individual firm which is based on asymmetric information between the entrepreneur and her investor.

2.1.1 Modigliani Miller proposition

In their fundamental article on "The cost of capital, corporation finance, and the theory of investment", Modigliani and Miller (1958, thereafter "MM") investigate whether it is less costly to finance an uncertain investment project with bonds, i.e. corporate debt, or with equity. In their famous Proposition 1 they show that *"the market value of any firm is independent of its capital structure"*. The necessary conditions to obtain this result are as follows:

1. All bonds yield a constant and certain income per unit of time.
2. Bonds and stocks are traded in perfect market. Firms are sorted according to their respective risk class. Securities of the same risk class are exchanged at the same price. Thus, the model builds heavily on arbitrage opportunities in the financial markets.
3. In the basic model, there is no tax shield for debt interest payments.

According to the MM model, no optimal capital structure exists – all structures being equivalent from the point of view of the cost of capital: A firm cannot reduce the cost of capital and increase its market value by financing part of its capital through bonds, even though debt money appears to be cheaper. The reason for this is that, as the firm increases its debt, the remaining equity becomes more risky. As this risk rises, the cost of equity rises as a result. The increase in the cost of equity capital offsets the higher proportion of the firm financed by low-cost debt. In fact, MM prove that the two effects exactly offset each other so that both the value of the firm and the firm's overall cost of capital are invariant to leverage.

The conditions for the "irrelevance" theorem were later simplified to the assumptions of perfectly competitive capital markets with symmetric information among all agents, no transaction costs, infinitely divisible securities, and no taxes. Moreover, the result hinges on perfect arbitrage and on the assumption that individuals can borrow at the same interest rate as firms.

Gottardi (1995) shows that the irrelevance result originates from a fundamental linearity property of the problem, independent of the completeness of the market. The MM result implies that the return on equity changes linearly when the firm's capital structure is modified. By contrast, the presence of a derivative security in capital restructuring will modify the value of the firm. The reason is that the payoff of derivative securities is affected in a non-linear way by changes in the financial structure. Thus, the financial decisions in this case do matter for the firm's market value.

2.1.2 Theories of optimal capital structure

Additional literature on the non-irrelevance of capital structure focuses on tax benefits of debt, agency costs, asymmetric information, and corporate control considerations. These four categories can be described as follows:

1. Trade-off theory

   The MM proposition is most unrealistic about taxes (cf. Modigliani and Miller 1963; or Modigliani 1988). If a company has taxable income, an increase in debt will reduce taxes paid by the company. The trade-off theory argues that value-maximizing firms attain an optimal capital structure by comparing the advantages of the interest tax shield of debt to the aggregated costs of financial distress. Costs of financial distress include costs of bankruptcy as well as the subtler monitoring and contracting costs which can erode the firm's value even if formal default is avoided.

2. Agency costs

   The agency cost approach states that a firm's capital structure is used to ameliorate conflicts of interest among various groups of claimants to the firm, including managers (see Harris and Raviv 1991 for a survey). When a firm has debt, selfish stockholders may pursue strategies which lower the market value of the firm and are less favorable for bondholders. The agency costs of equity may include shirking and perquisites, e.g. a big office, a company car, expense-account travels. Debt will help to reduce some of these agency costs: As pointed out by Jensen (1986), debt commits the firm to pay out cash. Thus, debt reduces the amount of "free cash flow" available to managers to engage in inefficient activities. This mitigation of conflict between managers and equity holders constitutes the benefit of debt financing.

On the other hand, conflicts of interest arise because the debt contract gives equity holders an incentive to invest in very risky projects. However, if debt holders anticipate this behavior, equity holders will receive less for the debt. This effect is called the "asset substitution effect" and represents an agency cost of debt financing. Jensen and Meckling (1976) argue that an optimal capital structure is obtained by trading off the agency cost of debt against the benefit of debt.

3. Asymmetric information

   To convey private information to capital markets or to mitigate adverse selection effects, a firm's capital structure may be used as a signaling device (Ross 1977, Leland and Pyle 1977, Poitevin 1989 in (2.2.3) below). Moreover, asymmetric information between managers and shareholders may induce a ranking of financial instruments, which is known as the financial pecking-order hypothesis (Myers 1984, Myers and Majluf 1984). The financial pecking order states, that in case of superior information of the management, firms will prefer to finance new investment opportunities first with internal funds, then with external debt, and only as a last resort with external equity, i.e. shares (for more details see (4.1.1) below).

4. Corporate control

   This theory of capital structure is driven by corporate control considerations. Harris and Raviv (1989) for example show how a firm's capital structure is used to allocate control rights between various claimants to the firm's profits. In (4.2.1) below, we will discuss how financial instruments based on debt and equity are designed to allocate control rights in venture capital contracting.

In the subsequent chapters we neglect tax issues. Instead, our analysis focuses on capital structure decisions which are influenced by agency costs, asymmetric information, and corporate control aspects. In the next section, we turn to the industry level and investigate how the firm's financial structure influences strategic competition in the product market.

2.2 Financial structure and product market competition

> *[O]utput market behavior will, in general, be affected by financial structure. [...F]oresighted firms will anticipate output market consequences of financial decisions; therefore, output market conditions will influence financial decisions.*
>
> Brander and Lewis (1986)

Corporate finance models and the theory of industrial organization have long been two separate fields of economic research. On the one hand, corporate finance literature after the Modigliani Miller proposition has concentrated on the different financial instruments for uncertain investment projects and their impact on the firm's market value. The theory of industrial organization, on the other hand, has focused on the strategic interaction of firms in the product market, and did not take interactions between product and input markets into account.

Only very recently, researchers have started to combine financing decisions of firms in the capital market with strategic competition in the product market. If financial markets are imperfect due to market power or asymmetric information, these financial market imperfections will lead to a reduction or a reinforcement of imperfections in the product market. In this chapter we will investigate the impact of capital structure decisions on firm behavior in the product market. We analyze how the financial structure decision affects investment strategies and the competitive behavior of firms in the product market, and how it finally determines the market structure of the industry.

As it is state of the art in industrial organization, the interaction between financial structure and product market competition is analyzed in a multi-stage game.

2.2.1 Optimal capital structure in a static framework

The analysis begins with the short-term impact of capital structure decisions on output strategies in a duopoly market. We present the model of Showalter (1995) who investigates the role of strategic debt for the price competition between two firms. Showalter shows that the limited liability effect of strategic debt induces firms to charge higher product prices in equilibrium. The result, however, is sensitive to the type of uncertainty the firms are confronted with.

Second, we investigate the impact of strategic debt on price and capacity competition between firms. Our model is a modified version of the Brander and Lewis (1986) approach. We show that if firms do not completely self-finance their investments, but take up debt for strategic reasons, firms will install larger capacities.

2.2.1.2 Limited liability effect and price competition

Showalter (1995) considers the choice of optimal strategic debt when firms compete in prices. The crucial assumption in his analysis is that firms which take up debt are protected by limited liability. Showalter shows that the result is sensitive to the type of uncertainty that exists in that market: He differentiates between cost and demand uncertainty. Showalter demonstrates that under price competition and demand uncertainty, firms will strategically invest in debt, while under cost uncertainty, firms will not take up any debt.

Assumptions

A1: Two symmetric firms i, j compete in prices with differentiated products.

A2: The time horizon is one period, $t=1$. The competition game consists of two steps: First, firms choose their optimal debt level, then they determine their optimal pricing strategy.

A3: The information structure is such that the production technology and the debt levels are common knowledge, i.e. they are observable and verifiable. The random variable z influences either the firm-specific demand or the marginal costs *after* debt levels and pricing strategies are chosen. The random variables are distributed over the interval $z_i, z_j \in [\underline{z}; \overline{z}]$ and have the density function $\varphi(z)$ and the distribution function $\Phi(z)$.

A4: All players are risk-neutral.

A5: Production technology: Each firm produces with zero fixed, but constant marginal costs. The profit function of firm i depends on its own price, its rival's price and the state of nature: $\Pi_i(p_i, p_j, z_i)$, with $\partial^2 \Pi_i / \partial p_i^2 < 0$, $\partial \Pi_i / \partial p_j > 0$, $\partial^2 \Pi_i / \partial p_i \partial p_j > 0$, and for firm j respectively. Higher z_i-values imply better states of nature and will lead to higher operating profits. Moreover, it is assumed that marginal profits are higher in better states of the world, $(\partial \Pi_i / \partial p_i) / \partial z_i > 0$.

A6: The firm specific demand is decreasing in its own price, $\partial q_i / \partial p_i < 0$, and increasing in the rival's price, $\partial q_i / \partial p_j > 0$.

The model

We consider a two-stage duopoly game, in which firms first choose their debt obligation and then determine the pricing strategy for their substitutive products. Equity holders, i.e. the owners of the firm, are subject to limited liability: In case the firm cannot meet its debt obligation, the firm goes bankrupt, and the debt holders have to bear the losses. Therefore, in choosing the optimal pricing strategy, the owners of the firm maximize the equity value of the firm, which is given by operating profits Π_i in good states of nature $z_i \in [\hat{z}_i, \overline{z}]$ minus the debt repayment D_i:

$$\max_{p_i} V_i(p_i, p_j, z_i) = \int_{\hat{z}_i}^{\overline{z}} (\Pi_i(p_i, p_j, z_i) - D_i) \varphi(z_i) dz_i \qquad (2.1)$$

How does the strategic debt influence the pricing strategy of each firm? This depends on the kind of uncertainty prevalent in that market. In case of cost uncertainty, operating profits are given as

$$\Pi_i(p_i, p_j, z_i) = (p_i - [c_i - z_i]) \cdot q_i(p_i, p_j), \qquad (2.2a)$$

where p_i stands for the product price, q_i for the firm-specific demand, and c_i for marginal production costs. Higher realizations of z_i imply better states of the world, i.e. lower marginal costs.

In case of uncertain demand, operating profits are given by

$$\Pi_i(p_i, p_j, z_i) = (p_i - c_i) \cdot [q_i(p_i, p_j) + z_i]. \qquad (2.2b)$$

The break-even point $\hat{z}_i$ separates the region where equity holders are residual claimants to the firm's returns (favorable states of nature) from the region where debt holders are residual claimants (unfavorable states).

$$\Pi_i(p_i, p_j, \hat{z}_i) - D_i = 0, \qquad (2.3)$$

In order to derive in which way prices and debt levels influence the break-even point, we totally differentiate equation (2.3) with respect to D_i, D_j, p_i, p_j. We obtain

- An increase in the debt level shifts the break-even point to the right:

$$d\hat{z}_i / dD_i > 0. \qquad (2.4a)$$

  If debt obligation increases, the realization of the random variable has to take higher values such that positive profits are obtained.
- An increase in the rival's debt level has no influence on the break-even point:

$$d\hat{z}_i / dD_j = 0. \qquad (2.4b)$$

- An increase in the own product price shifts the critical value $\hat{z}_i$ to the left

$$d\hat{z}_i / dp_i < 0. \qquad (2.4c)$$

- An increase in the rival's price shifts the critical value of $\hat{z}_i$ likewise to the left

$$d\hat{z}_i / dp_j < 0. \qquad (2.4d)$$

We see that $\hat{z}_i$ will decrease if the own price or the rival's price increase. The break-even point does not react to an increase of the rival's debt, but is shifted to the right if the firm takes up a higher leverage.

Product market stage

The two-stage game is solved by backward induction. In the second stage, debt levels D_i, D_j are taken as given and the firms choose their pricing strategy to maximize the equity value of the firm. Showalter argues that equity holders no

2.2 Financial structure and product market competition

longer have an incentive to consider the objectives of the debt holders once the debt is lent to the firm. The first-order condition of equation (2.1) equals:

$$\frac{\partial V_i}{\partial p_i} = \int_{\hat{z}_i}^{\bar{z}} \frac{\partial \Pi_i(p_i, p_j, z_i)}{\partial p_i} \varphi(z_i) dz_i \stackrel{!}{=} 0. \tag{2.5}$$

We know from (2.4d) that an increase in debt obligation shifts the break-even point to the right, which, in turn, affects the size of the marginal equity value and, therefore, the optimal price level. To analyze this effect in detail, we have to distinguish between cost- and demand uncertainties.

In case of demand uncertainty, an increase in debt D_i downsizes the equity holders' relevant region. Due to the positive slope of the marginal profit curve – expected marginal profits will become positive in the new interval $z_i \in [\hat{z}_i', \bar{z}]$ (see Figure 2.1 below). This causes firm i to raise its product price and shifts the marginal profit curve to the right.

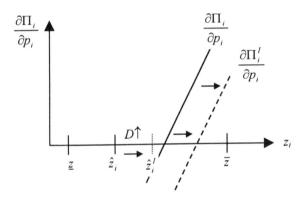

Figure 2.1: Effect of an increase in debt under demand uncertainty (Showalter 1995, 651)

In case of cost uncertainty, an increase in debt D_i also reduces the region of the equity holders' relevant states of nature. Due to the negative slope of the marginal profit curve – expected marginal profits will become negative in the new interval $z_i \in [\hat{z}_i', \bar{z}]$ (see Figure 2.2 below). This induces firm i to lower its optimal product price and shifts the marginal profit curve to the right.

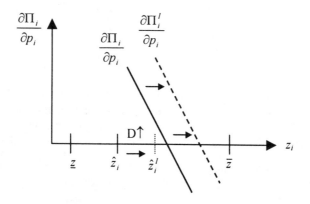

Figure 2.2: Effect of an increase in debt under cost uncertainty (Showalter 1995, 651)

When analyzing the impact of firm i's strategic debt D_i on the rival's pricing strategy, we have to take into account that prices are strategic complements and, therefore, move in the same direction. In case of demand uncertainty, taking up debt causes both prices, p_i and p_j, to rise. In case of cost uncertainty, however, taking up debt induces both prices p_i and p_j to fall.

Financial stage

With higher strategic leverage, interests of debt- and equity holders become more divergent, since equity holders bet on a smaller region of positive states of nature. Rational lenders anticipate that these competing interests will be internalized. Thus, when choosing the optimal debt level, the total firm value (and not only the equity value) will be maximized. The condition for the optimal leverage consists of:

$$\frac{\partial Y_i}{\partial D_i} = \left[\int_{\underline{z}}^{\hat{z}_i} \frac{\partial \Pi_i}{\partial p_i} \varphi(z_i) dz_i \right] \frac{dp_i}{dD_i} \quad \text{bankruptcy region}$$

$$+ \left[\int_{\hat{z}_i}^{\overline{z}} \frac{\partial \Pi_i}{\partial p_i} \varphi(z_i) dz_i \right] \frac{dp_i}{dD_i} \quad \text{success region} \qquad (2.6)$$

$$+ \left[\int_{\underline{z}}^{\overline{z}} \frac{\partial \Pi_i}{\partial p_j} \varphi(z_i) dz_i \right] \frac{dp_j}{dD_i} \quad \text{strategic effect of debt}$$

$$\overset{!}{=} 0.$$

2.2 Financial structure and product market competition

The first term describes the cost of an increase in debt D_i in the bankruptcy region. Since lenders obtain a repayment smaller than D_i in the bankruptcy region, this term is negative. The second term represents the impact of debt in the success region. This effect is zero due to the maximization condition of the subsequent pricing stage (envelope theorem). The last term reflects the strategic effect of debt, which is negative under cost uncertainty and positive under demand uncertainty. Taking all three effects together, we can state:

Proposition 2.1 (Price competition and strategic debt)

(i) *If demand is uncertain, firms in equilibrium will choose a positive amount of strategic debt, $D_i=D_j>0$. Product prices of these indebted firms are higher than under self-financing.*

(ii) *If costs are uncertain, firms will choose not to take up any strategic debt in equilibrium, $D_i=D_j=0$.*

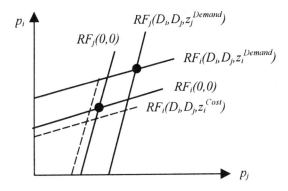

Figure 2.3: Product market equilibria if firms compete in prices and take up strategic debt (own presentation)

Discussion

If firms compete in prices and may take up debt for strategic reasons, the market results will depend on the type of uncertainty that exists. In case of price competition and demand uncertainty, taking up strategic debt leads to an increase in product prices and a reduction in output quantities. Competition, therefore, becomes less intense. As we will show below, this theoretical prediction fits very well with empirical findings.

By contrast, in case of price competition and cost uncertainty, firms restrain from taking up strategic debt, because the debt-induced equilibrium prices would be lower than under self-financing, which implies lower profits for the firms. Recall that in Showalter's model firms don't have any financing needs. It is only the limited liability effect which makes them choose to take up debt for strategic reasons.

Showalter provides an interesting analysis of price competition where the individual firm's debt-equity decision clearly influences the outcome in the product market. However, the model can be criticized for *(i)* its static approach and *(ii)* the rudimentary description of the investor-firm relationship. It is unclear whether the debt holders will actually obtain a non-negative return on investment (i.e. whether their participation constraint will be fulfilled or not). Moreover, since prices are typically adjusted very quickly once the demand or cost uncertainty is resolved, the commitment value of debt on a firm's pricing strategy might be overstated. In the next subsection, we, therefore, discuss the impact of strategic debt on quantity competition between firms.

2.2.1.2 Limited liability effect and capacity competition

In this subsection we investigate the impact of strategic debt on quantity competition. The basic work is the seminal paper of Brander and Lewis (1986). However, we present a modified version of their approach, since Brander and Lewis assume that firms compete in homogeneous quantities – which is quite unrealistic, given the fact that the vast majority of firms actually compete in prices. We therefore replace the original quantity competition by a two-stage game, in which firms first have to select their output capacities, before they engage in price competition. As Kreps and Scheinkman (1983) show, this capacity precommitment and its subsequent price competition yields the same outcome as the original Cournot game.

Prior to the capacity game, firms determine their financial structures. The key assumption here is again that firms which take up debt are protected by limited liability. As we will show below, this limited liability effect causes leveraged firms to behave more aggressively in the output market and induces them to install larger capacities.

Assumptions

A1: Players: There are two symmetric firms i, j, that compete in prices on a homogeneous market. Each firm has a rigid capacity constraint $\overline{K}_i, \overline{K}_j$. If firms decide to take up strategic debt, they will be protected by limited liability.

2.2 Financial structure and product market competition

A2: The time horizon is one period, $t=1$. The competition game consists of three steps: First, firms choose their optimal capital structure, then, second, they determine their optimal output capacity, before they finally select their product prices.

A3: Information structure: The price competition in the final stage takes place under full information. However, the individual firm's profits are initially uncertain because they are influenced by a random variable z_i. This random variable represents a firm-specific shock to the marginal cost of capacity: Higher z_i-values imply lower costs of capacity.[1] The realization of z_i takes place only *after* the debt levels and capacities have been chosen. The random variables are distributed over the interval $z_i, z_j \in [\underline{z}; \overline{z}]$; they have the density and distribution functions $\varphi(z_i)$ and $\Phi(z_i)$, respectively. Each firm's production technology and debt level are observable and verifiable by all players.

A4: Both firms as well as any potential investors are risk-neutral.

A5: Production technology: Firms choose the prices of their homogeneous products simultaneously. The marginal cost of production is assumed to be zero up to $\overline{K}_i$, and ∞ after $\overline{K}_j$ (for firm j respectively).

Previous to the price game, the capacity $\overline{K}_i$ must be installed at a unit cost of $[c_{K_i} - z_i]$. Note that the original capacity costs c_{K_i}, c_{K_j} of the two firms are symmetric. Firm i's operating profits are given as
$$\Pi_i(p_i, p_j, K_i, K_j, c_{K_i}, z_i) = p_i \cdot K_i - [c_{K_i} - z_i] K_i.$$
In reduced form, these profits are expressed as $\Pi_i(K_i, K_j, z_i)$, with $\partial^2 \Pi_i / \partial K_i^2 < 0$, $\partial \Pi_i / \partial K_j < 0$, $\partial^2 \Pi_i / \partial K_i \partial K_j < 0$. Moreover, it is assumed that marginal profits are higher in better states of the world, $(\partial \Pi_i / \partial K_i) / \partial z_i > 0$.

A6: Demand side: The inverse demand function $P(K_i, K_j)$ is linear, has a negative slope, and fulfills the concavity requirement.

As far as the financing side is concerned, firm i's capital structure is summarized by the variable D_i. D_i represents the debt obligation that a firm will have to repay at the end of the period if it decides to take up debt for strategic reasons.

The break-even point for which firm i can just repay its debt obligation out of current earnings is given by $\hat{z}_i$, assuming that $\underline{z} < \hat{z}_i < \overline{z}$:

---

[1] In contrast to the previous subsection, we neglect here the case of demand uncertainty, since capacity constraints are binding and the output cannot easily be adjusted to a higher or a lower demand quantity.

2 Financial structure and strategic competition

$$\Pi_i(K_i, K_j, \hat{z}_i) - D_i = 0. \tag{2.7}$$

Realizations above $\hat{z}_i$ result in positive net returns, while realizations below $\hat{z}_i$ imply net losses. In the first case, the firm's equity holders are the residual claimants of the profits. In the latter case, debt holders are the residual claimants, i.e. all gross profits are transferred to the investor and serve as a partial debt repayment.

This break-even point $\hat{z}_i$ depends implicitly on the variables K_i, K_j and D_i. We take the total differential of (2.7) to obtain the determinants for the level of $\hat{z}_i$:

- An increase in the debt level shifts the break-even point to the right:

$$d\hat{z}_i / dD_i = 1/(\partial \Pi_i / \partial z_i) > 0. \tag{2.8a}$$

If debt obligation increases, the realization of the random variable has to take higher values such that positive profits are obtained.
- An increase of the rival firm's debt has no influence on the break-even point:

$$d\hat{z}_i / dD_j = 0. \tag{2.8b}$$

- An increase in output quantity has no clear impact on the critical value $\hat{z}_i$:

$$d\hat{z}_i / dK_i = -\frac{\partial \Pi_i / \partial K_i}{\partial \Pi_i / \partial z_i}, \tag{2.8c}$$

since marginal profits from an additional unit of capacity $\partial \Pi_i / \partial K_i$ can either be greater than, equal to, or smaller than zero.
- An increase in the rival's capacity, however, shifts the critical value of $\hat{z}_i$ to the right, since the marginal profits from an additional unit of the rival's capacity $\partial \Pi_i / \partial K_j$ are definitively negative:

$$d\hat{z}_i / dK_j = -\frac{\partial \Pi_i / K_j}{\partial \Pi_i / \partial z_i}. \tag{2.8d}$$

Thus, if the rival firm increases its capacity K_j, then for a given capacity K_i the random variable z_i has to realize higher values for net profits to be positive.

Price competition stage

We solve the game by backward induction and begin with the price competition stage. On this stage, optimal prices are chosen while output capacities and debt levels are taken as given.

2.2 Financial structure and product market competition

Since both firms produce homogenous goods, their product prices will be identical, $p_i = p_j = p^*$. The market clearing price is determined via the demand function, after both firms have supplied their capacities to the market,

$$p^* = P(\overline{K}_i, \overline{K}_j).$$

Capacity competition stage

Stepping back one stage, we solve for the optimal capacity. At this stage, the debt levels D_i, D_j are taken as given, but the random variable z_i has not been realized yet. Since equity owners of the firm are subject to limited liability and do not have to personally make up for losses, they will choose for given debt levels D_i, D_j, an output strategy that maximizes the expected equity value of the firm. This means that they consider the firm value only in good states of nature, $z_i \geq \hat{z}_i$:

$$\max_{K_i} V_i = \int_{\hat{z}_i}^{\bar{z}} (\Pi_i(K_i, K_j, z_i) - D_i) \cdot \varphi(z) dz. \tag{2.9}$$

The necessary condition for maximizing the firm's equity value over output capacity is:

$$\frac{\partial V_i}{\partial K_i} = \int_{\hat{z}_i}^{\bar{z}} \frac{\partial \Pi_i(K_i, K_j, z_i)}{\partial K_i} \varphi(z_i) dz_i \stackrel{!}{=} 0 \tag{2.10}$$

The sufficiency condition $\partial^2 V_i / \partial K_i^2 < 0$ shall be fulfilled. Since the marginal equity value decreases with an increasing rival's capacity, $(\partial V_i / \partial K_i)/\partial K_j < 0$, the reaction function $K_i(K_j)$ has a negative slope, i.e. the firms compete in strategic substitutes. Moreover, for stable reaction functions and the existence of an unique equilibrium we require that the determinant of the Hesse-matrix is positive.

Recall that marginal profits from an additional unit of capacity increase with higher realizations of z_i, $(\partial \Pi_i / \partial K_i)/\partial z_i > 0$. An increase in the firm's debt level D_i shifts the break-even point $\hat{z}_i$ to the right. At the same time, the marginal profit curve also shifts to the right. This implies that the maximum of the firm's equity value is attained at a higher capacity only. The equity owners of the firm, therefore, will switch to a more aggressive output strategy and will install a higher capacity than under self-financing.

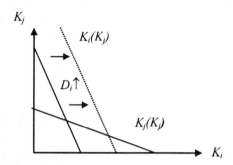

Figure 2.4: Outward shift of the capacity reaction function due to an increase of debt (adapted from Brander and Lewis 1986, 964)

Since the capacity choice is made before the uncertainty about z_i is resolved, equity holders will, after taking up a higher amount of debt, speculate on a higher realization of z_i. This scenario corresponds to the findings of Myers (1977), who states that if the debt-equity ratio of a firm increases, the management of the firm will select riskier investment strategies, since they bet on positive state of nature realizations and take only positive profits into consideration.

Proposition 2.2 (Strategic debt and capacity competition)

If debt is taken up for strategic reasons, the firm's equity holders will switch to a more aggressive output strategy and will install higher capacities.

Corollary 2.1

An unilateral increase in firm i's debt D_i causes an increase in firm i's capacity K_i and a decrease in the rival's capacity, K_j.

Financial stage

Rational lenders anticipate that a higher leverage commits equity holders to a more aggressive output strategy which, in turn, increases the risk of bankruptcy. Debt holders, therefore, require that this conflict of interest between debt and equity holders must be internalized. Thus, when choosing the optimal debt level, the total firm value - which consists of the equity value V_i and the debt values W_i - is maximized:

2.2 Financial structure and product market competition

$$\max_{K_i} Y_i = V_i + W_i = \int_{\hat{z}_i}^{\bar{z}} (\Pi_i(K_i, K_j, z_i) - D_i) \cdot \varphi(z_i) dz_i$$

$$+ D_i [1 - \Phi(\hat{z}_i)] + \int_{\underline{z}_i}^{\hat{z}_i} \Pi_i(K_i, K_j, z_i) \cdot \varphi(z_i) dz_i \quad (2.11)$$

$$= \int_{\hat{z}_i}^{\bar{z}} \Pi_i(K_i, K_j, z_i) + \int_{\underline{z}_i}^{\hat{z}_i} \Pi_i(K_i, K_j, z_i) \cdot \varphi(z_i) dz_i$$

Note that in the last line we take into account that $\int_{\hat{z}_i}^{\bar{z}_i} D_i \varphi(z_i) dz_i = D_i [1 - \Phi(\hat{z}_i)]$.

To optimize the total firm value is to maximize the expected profits over all states of the world.

We now investigate how an increase in debt level D_i affects the total value of firm i. We state again that issuing debt in this framework is purely for strategic reasons, the firm itself doesn't have any external financing need. By differentiating the total firm value over the debt level D_i, we obtain:

$$\frac{dY_i}{dD_i} = \left[\int_{\underline{z}}^{\hat{z}_i} \frac{\partial \Pi_i}{\partial K_i} \varphi(z_i) dz_i \right] \frac{dK_i}{dD_i} \qquad \text{bankruptcy region}$$

$$+ \left[\int_{\hat{z}_i}^{\bar{z}} \frac{\partial \Pi_i}{\partial K_i} \varphi(z_i) dz_i \right] \frac{dK_i}{dD_i} \qquad \text{success region} \quad (2.12)$$

$$+ \left[\int_{\underline{z}}^{\hat{z}_i} \frac{\partial \Pi_i}{\partial K_j} \varphi(z_i) dz_i + \int_{\hat{z}_i}^{\bar{z}} \frac{\partial \Pi_i}{\partial K_j} \varphi(z_i) dz_i \right] \frac{dK_j}{dD_i} \quad \text{strategic effect of debt}$$

The first term represents the impact of a higher debt level on the bankruptcy region ($z_i < \hat{z}_i$). Under the assumption $(\partial \Pi_i / \partial K_i)/\partial z_i > 0$, an increase in D_i leads to an increase in optimal output capacity, which, at the same time, leads to higher losses for a given z_i value in the bankruptcy region. Thus, taking on more debt exacerbates the conflict of interest between debt and equity holders. This implies that the first term has a negative impact on Y_i. The second term is equal to zero due to the first order condition under equity value maximization (2.10). Finally, the third term represents the strategic effect of debt. It shows the impact of a higher debt level D_i on equilibrium capacity of firm j: Since firm i will increase its capacity K_i after an increase of D_i, rival j is forced to decrease its capacity according to the negative slope of the reaction function. The strategic effect of debt raises both the debt and equity values of the firm.

Taking all three effects together, the positive strategic effect dominates the negative effect from the conflict of interest for sufficiently small levels of debt.

Especially for an initial debt level of $D_i=0$, the marginal effect from an increase in debt on total firm value is strictly positive, $dY_i / dD_i > 0$.

According to the investment strategy classification of Fudenberg and Tirole (1989), this results in a "top-dog" strategy, which means that it is favorable for each firm to strategically invest in debt and, thus, to commit to an aggressive output strategy. If firms choose a positive debt level, they will install higher output capacities than in the standard Cournot oligopoly where firms are completely self-financed.

Proposition 2.3 (Capital structure and capacity competition)

If leveraged firms are protected by limited liability and firms compete in capacities, both firms will take up strategic debt, $D_i>0$, $D_j>0$, in equilibrium.

This strategy, however, has serious implications on the joint profits of both firms: If both firms choose higher capacities, the value of the industry will not be maximized. Instead, the firms are confronted with a prisoner's dilemma: Joint profits are highest when both firms produce under complete self-financing. This, however, is not a dominant strategy, because it pays for each firm to unilaterally deviate from the all-equity equilibrium. From Corollary 2.1 we know that a firm which unilaterally takes up debt will realize higher profits that its self-financed rival. In equilibrium, both firms therefore choose to issue debt and install higher capacities. This finally results in lower profits, $\Pi(D_i,D_j)<\Pi(0,0)$:

Firm i / Firm j	$D_j = 0$	$D_j > 0$
$D_i =0$	$\Pi_i(0,0) = \Pi_j(0,0)$	$\Pi_i(0,D_j) < \Pi_j(0,D_j)$
$D_i > 0$	$\Pi_i(D_i,0) > \Pi_j(D_i,0)$	$\Pi_i(D_i,D_j) = \Pi_j(D_i,D_j)$

Table 2.1: Payoffs according to the firms' respective debt levels

2.2 Financial structure and product market competition   21

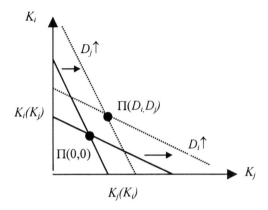

Figure 2.5: Output market equilibrium with and without firms taking up debt
(own presentation)

Proposition 2.4 (Capital structure, capacity competition, and profits)

If firms compete in capacities and take up debt for strategic reasons, profits are lower than under pure equity financing. The firms are trapped in the prisoner's dilemma.

Discussion

The combined Brander and Lewis (1986) and Kreps and Scheinkman (1983) model shows that leverage and the associated limited liability effect commits firms to install higher output capacities and induces them to behave more aggressively in the output market. Taking up debt for strategic reasons serves to weaken the rival's competitive position.

Since both firms in the model behave non-cooperatively, the final outcome results in lower profits. Firms find themselves trapped in the so-called prisoners' dilemma. However, firms could improve their profit situation, especially if there are only very few of them in the market, by coordinating their capacities. We admit, though, that collusion is forbidden in most industries by the respective antitrust laws. And the history of OPEC shows, that capacity cartels are not easy to sustain.

Nevertheless, dynamic aspects of product market competition can only rudimentarily be captured in this static model. It would be better, therefore, to extend the one-period framework to a two-period model, in which firms can explicitly react to their rival's strategy and take intertemporal aspects into account.

As far as the financing side is concerned, the financial relationship between the firm and the outside investor is sketched only very briefly. It is important to stress that D represents not the debt level, but the amount of debt repayment. Nothing is said about the initial amount of funds provided by the lenders. And no information is provided on how the firm actually spends this money. The model, therefore, is not really closed. In chapters 3 and 4, we take this criticism into account and examine the financial contracts between firm and its investors more in detail.

2.2.2 Capital structure in a dynamic framework

In order to circumvent the missing dynamics in the models presented above, we now will investigate the capital structure and product market interaction if firms compete twice on the product market. First, we study dynamic price competition in a very general case. Then we successively include consumer switching costs (2.2.2.2), product quality decisions (2.2.2.3) and predatory pricing (2.2.2.4) into our analysis. This allows us to emphasize different aspects of financial policy and pricing strategy decisions and to analyze the capital structure choice of firms under various forms of product market competition.

2.2.2.1 *Price competition and long-term debt*

Dasgupta and Titman (1998) analyze product market and financial market interaction in a two-period model of price competition where firms take up long-term debt.

Assumptions

A1: Players: Two symmetric firms i, j produce similar, but differentiated products and compete in prices in the product market.

A2: The time horizon is two periods, $t=1,2$. In each period, firms have to make a capital structure decision, before they select the optimal product prices for that period.

A3: Information structure: The only source of uncertainty in this model is a random liquidation value. Both firms are liquidated at the end of the second period, no matter whether they are solvent or not. The random liquidation value generates an additional cash flow, it lays in the interval $z \in [\underline{z}; \bar{z}]$ (Dasgupta and Titman actually set the lower limit to zero), and has a continuous density $\varphi(z)$ and distribution function $\Phi(z)$. Each firm's production technology is common knowledge, and the levels of debt are observable and verifiable by all parties.

2.2 Financial structure and product market competition

A4: All players are risk-neutral.

A5: Production technology: Initially, each firm produces with zero fixed and constant marginal costs. At the beginning of the second period, however, an additional investment I_i is required, which exceeds the first-period profits. First-period profits depend on first-period prices, $\Pi_i^1(p_i^1, p_j^1)$, while second-period profits depend on second-period prices and on the market share attracted in the previous period, σ_i and σ_j (with $\sigma_j = 1 - \sigma_i$). Since the second-period prices are functions of the first-period market shares, we can write $\Pi_i^2(\sigma_i)$ for the profits in the second period. Moreover, we assume that second-period profits increase as the customer base increases, $\partial \Pi_i^2 / \partial \sigma_i > 0$, and that this market share negatively depends on first-period prices, $\partial \sigma_i / \partial p_i^1 < 0$.

A6: Demand side: Consumers have exogenous tastes for the product variants supplied. They are sensitive to price, but tend to favor the firm from which they purchased the product in the previous period.

A7: Financial policy: Each firm decides upon taking up long-term debt at the beginning of the game. Total repayment D_i is due at the end of the second period. In addition to this, firms need to finance the intermediate investment I_i at the end of period 1. In case the investment I_i is (partly) financed with debt, this intermediate debt is junior to the long-term debt D_i.

The model

If the firm is completely self-financed, i.e. $D_i=0$, the firm value is characterized by the sum of first- and second period profits, plus the expected liquidation value, minus the fixed intermediate investment:

$$V_i = \Pi_i^1(p_i^1, p_j^1) + \Pi_i^2(\sigma_i(p_i^1, p_j^1)) + E(z_i) - I_i. \tag{2.13}$$

To choose the optimal first-period price is a present value problem since a marginal price reduction in the first period lowers the first-period profits, but increases the firm's market share and, thus, the second-period profits.

If firms take up long-term debt at the beginning of the game, the equity owners are subject to limited liability and will consider only solvent states of nature. The firm's equity value under debt financing is given by

$$V_i(D_i, D_j) = \Pi_i^1(p_i^1, p_j^1) - I + \int_{D_i - \Pi_i^2}^{\bar{z}} [\Pi_i^2(\sigma_i(p_i^1, p_j^1)) + z_i - D_i] \varphi(z_i) dz_i, \tag{2.14}$$

where the lower-limit of integration $D_i - \Pi_i^2$ is equivalent to the break-even point $\hat{z}_i$ of the previous models. The (gross) debt value for the long-term investors is given by:

$$W_i = D_i[1-\Phi(D_i - \Pi_i^2)] + \int_{\underline{z}}^{D_i - \Pi_i^2} (\Pi_i^2 + z_i)\varphi(z_i)dz_i, \qquad (2.15)$$

i.e. the long-term debt holders will obtain a fixed repayment D_i if the firm is solvent, and the remaining profits in case of failure. Total firm value if firms take up debt is derived by adding up equations (2.15) and (2.16), which results in:

$$Y_i = \Pi_i^1(p_i^1, p_j^1) - I_i + \int_{D_i - \Pi_i^1}^{\bar{z}} [\Pi_i^2(\sigma_i(p_i^1, p_j^1)) + z_i - D_i]\varphi(z_i)dz_i. \qquad (2.16)$$

The total firm value consists of first-period profits, minus the intermediate investment expenditure, plus the expected net profits from the second period.

Second-period pricing stage

We now analyze how taking up long-term debt will influence the firm's pricing strategies in both periods. Note that in Dasgupta and Titman's (1998) model, marginal profits are not directly affected by the random variable z_i.

To analyze the equilibrium pricing strategies, we begin as usual with the last period. For given market shares, σ_i and σ_j, second-period output prices are selected such as to maximize the firm's second-period profits. The first-order condition for this optimization problem is given as

$$\frac{\partial \Pi_i^2(p_i^2, p_j^2, \sigma_i)}{\partial p_i^2} \stackrel{!}{=} 0, \qquad (2.17)$$

and analogously for firm *j*. Thus, the maximization problem is exactly the same as under self-financing.

Second-period financial stage

As far as the financial stage of the second period is concerned, firms do not really decide upon their financial structure. Dasgupta and Titman rather *assume* that firms use their first-period profits to partly finance the intermediate investment I_i, such that a borrowing need of $I_i - \Pi_i^1$ is given.

First-period pricing stage

We therefore can immediately proceed to the first-period output stage. Differentiating equation (2.16) with respect to p_i^1, we obtain

$$\frac{\partial Y_i}{\partial p_i^1} = \frac{\partial \Pi_i^1(p_i^1, p_j^1)}{\partial p_i^1} + \frac{\partial \Pi_i^2}{\partial \sigma_i} \frac{\partial \sigma_i}{\partial p_i^1} \cdot [1 - \Phi(D_i - \Pi_i^2)] \stackrel{!}{=} 0. \quad (2.18)$$

The second term indicates that when optimizing the first-period pricing strategy, only solvent states of nature are taken into account. We suppose that the second-order condition $\partial^2 Y_i / \partial p_i^1 \partial p_i^1 < 0$ is fulfilled.

First-period financial stage

If a firm's level of long-term debt increases, the second term of (2.18) will decline, inducing marginal profits to fall. The firm will, ceteris paribus, increase its first-period price, causing the reaction function to shift outward. Moreover, as we see in Figure 2.6, the reaction function will also change its slope.

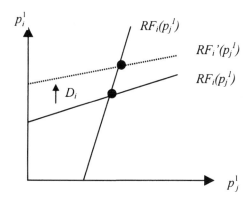

Figure 2.6: The effect of long-term debt on first-period prices
(Dasgupta and Titman, 1996, 23)

This change in slope is due to an additional negative effect, that we obtain by differentiating equation (2.19) with respect to the rival's price:[2]

---

[2] Dasgupta and Titman (1998, 713) have a small mistake in their derivative, which, however, does not change the qualitative result.

$$\frac{\partial^2 Y_i}{\partial p_i^1 \partial p_j^1} = \frac{\partial^2 \Pi_i^1}{\partial p_i^1 \partial p_j^1} + \frac{\partial \Pi_i^2}{\partial \sigma_i} \frac{\partial \sigma_i}{\partial p_i^1} \frac{\partial \Pi_i^2}{\partial p_j^1}[1 - \Phi(D_i - \Pi_i^2)] - \frac{\partial \Pi_i^2}{\partial \sigma_i} \frac{\partial \sigma_i}{\partial p_i^1} \frac{\partial \Phi(D_i - \Pi_i^2)}{\partial p_j^1}$$
(2.19)

The third term in equation (2.19) causes $\partial^2 Y_i / \partial p_i^1 \partial p_j^1$ to decline and the reaction function becomes flatter. The intuition behind this is that if firm i increases its debt level D_i, its marginal profits will decline, and firm i raises its price, which in turn induces firm j to raise its price (via the shift of RF_i). The rival's price increase, however, has a secondary positive effect on the profits of firm i which enlarges the respective success region. Therefore, firm i will slightly lower the first-period price in order to attract a higher market share. Or, as Dasgupta and Titman interpret it, the first-period pricing decision is the solution to a present value problem: Preexisting long-term debt raises the cost of new borrowing and thus reduces the incentive to gain market share by lowering price. If firm i takes up risky long-term debt, this will induce both firms i, j to raise their first-period prices. A price increase p_j^1 makes firm i more profitable. This, in turn, causes firm i's borrowing costs to decline, making firm i want to slightly lower its price again in order to gain market shares. This alters the slope of firm i's reaction function.

Proposition 2.5 (Pricing strategy and long-term debt)

(i) If firm i takes up long-term debt and requires additional junior debt to finance the intermediate investment, its first-period price increases in the level of existing debt.

(ii) Firm j's first-period price is also increasing in firm i's existing debt level.

(iii) If firm j is self-financed, while firm i unilaterally increases its long-term debt, firm j's first-period price will be lower than firm i's price.

Since the same reasoning holds for firm j, equilibrium product market prices will be higher for leveraged firms than for self-financed firms. At the same time, output quantities in leveraged industries are lower. Thus, long-term debt reduces the intensity of first-period price competition, while competition in the second period is intensified. In equilibrium, both firms will include some debt in their capital structures.

The first-period result of Dasgupta and Titman corresponds to the findings of Showalter for the static case: If firms take up debt, product prices will increase.

2.2.2.2 *Impact of consumer switching costs*

In order to derive a closed-form solution for the optimal debt levels as well as for the equilibrium prices and market shares, we have to further specify the demand side of the industry. Dasgupta and Titman (1998) integrate a model of consumer switching costs based on Klemperer (1997, 1995) into their model of product differentiation and price competition. As we have already noted in assumption A6, consumers have different preferences for the product variants supplied.

We assume that consumers are uniformly distributed along a line segment [0,1] ("Hotelling-Street"), with firms i and j located at both ends of this line. Consumers buy one unit of the good per period of time. In addition to the product price, they have to bear transportation costs T depending on the distance from the seller.

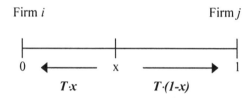

Figure 2.7: Hotelling line (adapted from Tirole 1988)

A consumer located at point x buys from firm i if the sum of product price and transportation costs is lower than when buying form firm j:

$$p_i^1 + Tx \leq p_j^1 + T(1-x),$$

$$x \leq \frac{T + p_j^1 - p_i^1}{2T}.$$

Hence, the market share for firm i amounts to

$$\sigma_i = 1/2 + \frac{p_j^1 - p_i^1}{2T} \qquad (2.20)$$

In the second period, consumers will buy from the same firm as in the previous period, if switching from product variant i to variant j (and vice versa) is associated with high costs. Especially if these switching costs exceed one half of the consumer's reservation price p^r,[3] then first-period market shares stay constant

---

[3] The reservation price p^r is the price at which consumers are just indifferent between buying the good or not.

2 Financial structure and strategic competition

in the second period, and the unique second period equilibrium is one where each firm charges the reservation price p^r.

Reference case: self-financed firms

Firm value in the all-equity case ($D_i=0$) consists of

$$V_i = p_i^1 \frac{T + p_j^1 - p_i^1}{2T} + p^r \frac{T + p_j^1 - p_i^1}{2T} + E(z_i) - I, \qquad (2.21)$$

i.e. the firm value sums up first and second period profits, and the expected liquidation value, and is reduced by the intermediate investment costs. As Klemperer (1993) shows, this switching cost approach has the standard properties of stable and upward sloping price reaction functions.

Optimal level of long-term debt

The optimal debt level is derived by differentiating total firm value from (2.16) with respect to long-term debt:

$$\frac{\partial Y_i}{\partial D_i} = \left[\frac{\partial \Pi_i^1}{\partial p_i^1} + \frac{\partial \Pi_i^2}{\partial \sigma_i}\frac{\partial \sigma_i}{\partial p_i^1}\right] \cdot \frac{\partial p_i^1}{\partial D_i} + \left[\frac{\partial \Pi_i^1}{\partial p_j^1} + \frac{\partial \Pi_i^2}{\partial \sigma_i}\frac{\partial \sigma_i}{\partial p_j^1}\right] \cdot \frac{\partial p_j^1}{\partial D_i} \stackrel{!}{=} 0. \qquad (2.22)$$

In case of the consumer switching cost model this amounts to

$$\frac{\partial Y_i}{\partial D_i} = 0 + \frac{\partial \Pi_i^2}{\partial \sigma_i}\frac{\partial \sigma_i}{\partial p_i^1} \cdot \Phi(D_i - \Pi_i^2) + \left[\frac{\partial \Pi_i^1}{\partial p_j^1} + \frac{\partial \Pi_i^2}{\partial \sigma_i}\frac{\partial \sigma_i}{\partial p_j^1}\right] \cdot \frac{dRF_j(p_i^1)}{dp_i^1} \stackrel{!}{=} 0, \qquad (2.23)$$

where the first term is zero due to the envelope-theorem. If we now assume that the liquidation value is distributed uniformly between $[0, \bar{z}]$, we can write equation (2.23) as

$$\frac{\partial Y_i}{\partial D_i} = -\frac{p^r}{2T}\left[\frac{D_i - p^r/2}{\bar{z}}\right] + \left[\frac{p_i^1}{2T} + \frac{p^r}{2T}\right] \cdot RF_j^{'} \stackrel{!}{=} 0, \qquad (2.24)$$

where $RF_j^{'}$ denotes the slope of the rival's reaction function. This slope equals

$$RF_j^{'} = \frac{1 - \frac{(p^r)^2}{2T \cdot \bar{z}}}{2 - \frac{(p^r)^2}{2T \cdot \bar{z}}}. \qquad (2.25)$$

By inserting this into equation (2.24), rearranging terms and simplifying we obtain for the optimal level of debt in the symmetric equilibrium:

$$D_i = D_j = \frac{T}{p^r} \cdot \bar{z}. \tag{2.26}$$

Thus, the optimal debt level increases the greater the product differentiation in period one, i.e. the higher the transportation costs. Moreover, the debt level raises with a higher (upper limit of) liquidation value $\bar{z}$. This implies that in industries where assets have higher liquidation values or in which firms have more tangible assets, long-term debt levels will be higher.

Discussion

In this two-period model of price competition between leveraged firms, the demand side is integrated via transportation costs and consumer switching costs. This enables Dasgupta and Titman to derive explicitly the optimal amount of long-term debt in industry equilibrium and to determine the optimal capital structure.

Taking up debt implies that both firms will raise their product prices in the first period. The existence of switching costs further implies that second-period prices will be even higher and increase up to the consumers' reservation price. Dasgupta and Titman (1998), however, do not explain which empirical phenomenon refers to the consumer switching costs. Do they mean information and search costs, the additional time required to initiate new business relationships, or transaction costs that arise when purchasing contracts have to be rewritten? If consumers actually have preferences for product diversity, then the switching cost approach does not seem very appropriate to provide a dynamic version of the Hotelling model.

2.2.2.3 Product quality and long-term debt

In their (1996) approach, Dasgupta and Titman investigate not only the dynamic interaction between capital structure and pricing strategies, but also analyze how an ex-ante unobservable product quality choice interacts with the competitive environment and the capital structure decision. The model is an extension to the consumer switching cost model and integrates imperfect information about the product quality as an additional variable into the game. The analysis provides an explanation for the empirical phenomenon that leveraged buyouts sometimes result in lower rather than in higher product prices.

Assumptions

A1: Players: Two firms compete on the basis of prices in the product market.

A2: The time horizon is two periods with multiple stages (see below).

A3: Information structure: The firms produce goods of either high or low quality. While it is common knowledge that firm i produces a high-quality good, firm j can choose between supplying high or low quality.

Consumers observe the product's quality only with a lag of one period, i.e. they know the true quality in the second, but not in the first period of competition. However, the price that the firm charges and the financial structure provide information for the consumers about the incentives to produce a high-quality good. In period one, the consumer form rational beliefs about the product's quality based on price and the firm's financial structure.

In addition to this, there is uncertainty about the firm's liquidation value at the end of period two. To keep things simple, it is assumed that the liquidation value equals $\bar{z}$ with probability θ, and zero with probability $(1-\theta)$.

A4: Both firms and all investors are risk neutral.

A5: Production technology: Each firm has to make a fixed investment I_t at the beginning of each period $t=1,2$ in order to start production. Moreover, high-quality production requires an additional investment I^{HQ} during the first period, which allows to produce high quality goods in both periods.

A6: Firm-specific demand depends on *(i)* the consumers' preferences for the product variant supplied, *(ii)* on the product price and *(iii)* on the beliefs about the product's quality. Under complete information, i.e. after the uncertainty about the product's quality is resolved in period 2, consumers are willing to pay a reservation price p^r for high-quality products, and a fraction of this, i.e. χp^r with $\chi<1$, for low-quality products. Consumers buy one unit of the product per period of time.

A7: Financial structure: First, we assume that both firms are self-financed. Then, we suppose that firm j is financially restricted and needs external debt financing.

The time structure of this multi-stage game is as follows:

Before production starts,

- firm j will take up debt if necessary,
- both firms make the first-period investment I_1, and
- firm j selects its product quality (firm i always produces high quality).

In the first period,

- firm i and j select their product price.
- Based on the prices and the observed capital structure, consumers form beliefs about the product qualities and make their period 1 purchasing decisions.

2.2 Financial structure and product market competition   31

- Consumers buy one unit of the product at firm i or j, and the firms realize their first-period profits.
- Firms will finance the additional investment I^{HQ}, if they decide to engage in high-quality production.

In the second period,

- both firms make the second-period fixed investment I_2.
- Then, the uncertainty about the product quality is resolved.
- Firms select their product prices, products are sold and profits are realized.
- Debt has to be repaid.
- Firms are liquidated. The liquidation value is either low, $z = 0$, or high, $z = \bar{z}$.

Product quality when firms are self-financed

In this subsection we concentrate on the interaction between product quality and pricing strategy. Since quality is unobservable in the first period, firm j must consider how its first-period price affects the consumers' beliefs about the product quality. It is common knowledge that firm i produces high quality, and firm j and all consumers share the common conjecture that firm i will charge the price $E(p_i^1)$.[4]

Firm j determines its product quality choice as follows: Suppose that p_j^1 is the price for which consumers believe that firm j, too, produces high quality products. If for this price p_j^1 and expected rival's price $E(p_i^1)$ firm j's profits minus the quality expenditures I^{HQ} are higher than the profits under low-quality production, firm j will choose to produce high-quality products:

$$\Pi_j^1[E(p_i^1), p_j^1] + \Pi_j^{2\,HQ}[\sigma_j(E(p_i^1), p_j^1)] - I^{HQ} \\ > \Pi_j^1[E(p_i^1), p_j^1] + \Pi_j^{2\,LQ}[\sigma_j(E(p_i^1), p_j^1)]. \quad (2.27)$$

Here, the superscript HQ stands for high quality, and LQ stands for low quality. By inserting firm j's firm-specific demand derived from the Hotelling model into (2.27), we obtain the condition for high quality production:

$$p^r(1-\chi)\frac{T + E(p_i^1) - p_j^1}{2T} \geq I^{HQ}. \quad (2.28)$$

The high quality condition (2.28) depends on the reservation price for high and low quality goods, firm j's market share, and the required amount of quality

---

[4] In equilibrium, of course, price conjectures will have to fulfill themselves.

investment. Thus, if in the first period firm j selects a price smaller than the price for which (2.28) is binding, $p \leq p_j^1$, the market believes that product quality is high. On the other hand, if firm j charges a higher first-period price $p > p_j^1$, then consumers estimate that product quality is low.

The intuition behind this -- at first contrary -- statement is that lower first-period prices imply a higher second-period market share. If consumers observe a high product price in the first period, they will infer that firm j attempts to realize high present-period profits in exchange for low second-period profits in case of low quality. If instead the first-period price is relatively low, consumers know that the firm wants to attract a large customer base in order to realize high second-period profits by charging the full reservation price p^r for its high quality product.

A Bayesian equilibrium in the first period will be given if firm j produces high-quality goods and charges the price p_j^1, which maximizes firm value under the high-quality restriction (2.28). At the same time the rival i must set its price according to its reaction function, which, in turn, corresponds to the expected and the actual price, $p_i^1(p_j^1) = E(p_i^1) = p_i^1$. If this pricing condition is fulfilled, both firms credibly commit to offer high-quality products. The equilibrium itself is not affected by the unobservability of quality.

Product quality when one firm needs debt financing

Next, we analyze how the interaction between quality choice and pricing strategy is affected by the capital structure decision of the firm. We extend the setting above in that we assume that firm j, which has to decide whether to produce high or low quality, does not have sufficient internal funds and has to take up debt. The rival i continues to be self-financed and to produce high-quality products.

To make things interesting, we focus on a certain interval of firm j's debt level:

$$\Pi_j^{2HQ} + 0 < D_j < \Pi_j^{2LQ} + \bar{z}.$$

On the one hand, the debt level shall be higher than second-period profits under high quality production and a liquidation value of zero. On the other hand, the debt level shall be smaller than second-period profits under low quality production and a positive liquidation value. Thus, firm j will go bankrupt if the liquidation value equals zero. The bankruptcy probability is equal to $1-\theta$, and the probability for a high liquidation value is θ.

The high-quality condition under debt financing, i.e. the condition for firm j to produce high quality goods even after taking up debt, is given by

2.2 Financial structure and product market competition

$$\theta \cdot p'(1-\chi)\left[\frac{T+E(p_i^1)-p_j^1}{2T}\right] \geq I^{HQ}. \quad (2.28')$$

In order to fulfill this modified condition (2.28') firm j's first-period price has to be reduced below the one in the self-financing case (2.28). If, however, the success probability θ is very small, producing high quality and charging the price p_j^1 is not an equilibrium strategy any more, since firm j wants to switch to low-quality production as soon as (2.28') is no longer fulfilled. This, in turn, induces a price change of the rival firm according to its reaction function.

If, on the other hand, $\theta p'(1-\chi) < 2I^{HQ}$ holds, then there exists a unique equilibrium in which both firms produce high-quality products and charge lower first-period prices than in case of all-equity financing of firm j.

Proposition 2.6 (Quality choice and debt)

Suppose firm i is all self-financed, while firm j needs external debt financing. Then, if condition (2.28') is fulfilled, both firms will produce high quality and will choose lower first-period prices than under complete self-financing.

Thus, by integrating the quality decision into the financing and competition game, price competition will become more intense.

This proposition provides an explanation for the empirically observed phenomenon that product prices may sometimes fall after a leveraged buyout: If the bankruptcy risk rises due to the massive increase in leverage, higher market shares will be necessary to credibly offer high-quality products to the consumers. This increase in market share is achieved via a price-cut in the first period. If, however, the rival firm i is self-financed, it will compete more fiercely for market shares, which in turn will oblige firm j to reduce its first-period price more strongly.

Moreover, if both firms are financially restricted and need to take up debt, then for given success probabilities θ_i and θ_j, both firms will charge higher prices for their high-quality products compared to the situation where only one firm is leveraged.

Discussion

In this dynamic model of price competition, firm j is in need of external debt financing and does not take up debt merely for strategic reasons. The original two-period price competition is extended to a quality-signaling game: By charging low first-period prices, firms can credibly offer high quality products.

This result stands in contrast to the previous models of Dasgupta and Titman (1998) and Showalter (1995). It demonstrates that the impact of debt on pricing strategies is sensitive to the form of competition and the type of uncertainty that prevail in the product market.

Here, the random variable z_i affects the game only in the second period, and determines whether the firm will be successful or go bankrupt. The higher the liquidation value, the more likely that the firm produces high-quality goods.

2.2.2.4 Predatory pricing

While in the previous subsection, the leveraged firm j initiated the price cuts in order to signal quality, we here investigate whether the self-financed rival, too, has an incentive to engage in price cuts. The idea behind is that a self-financed firm can prey upon its debt-financed rival and can force it into low-quality production.

If firm j has a sufficiently high debt level, it will need to charge a high first-period price in order to generate sufficient profits and to keep the second-period refinancing requirement as low as possible. On the other hand, the firm will need a sufficiently high market share to signal its quality to consumers.

A self-financed rival could now commit to a price slightly lower than the equilibrium price. At this price, firm j will have to switch to low quality production, which leads to a higher market share for firm i. We derive the following comparative static results with respect to capital structure:

1. If both firms are self-financed, it depends on the level of necessary quality investment I^{HQ} whether predatory pricing is profitable or not. If the required quality investment I^{HQ} is drawn from the interval $[\frac{1}{4}p^r(1-\chi); \frac{1}{2}p^r(1-\chi)]$, then for values smaller than I^{HQ} it will not be optimal for firm i to lower its product price: Although a price reduction of firm i will force firm j into low quality production, which, in turn, increases firm i's profits, firm i has also incentives to increase the price and to directly increase its profits. Therefore, under complete self-financing, predatory price cutting does not represent an equilibrium strategy (Dasgupta and Titman 1996, 30p.)

2. If one firm is self-financed while the other firm is debt-financed, predatory pricing may be successful. Predation is likely if the rival's bankruptcy probability $1-\theta_j$ (due to a liquidation value of $z=0$) is relatively high. We know that firm j's profits increase in the success probability θ_j. Assume for a moment that this success probability is just above the critical value $\theta_j > \theta_j^o$, such that firm j weakly prefers high-quality production. If firm i now cuts its price for strategic reasons, firm j is forced to produce goods of low quality. Rival i thus obtains a higher market share, which leads to an increase in

second-period profits. Whether this predatory pricing is profitable or not depends on the price induced losses of first-period profits.

3. If both firms i and j are debt financed, then there exists a range of values for the rival's bankruptcy probability $1-\theta_j$, such that predation is no longer profitable for firm i. Empirical evidence suggests that incentives to prey will be reduced after a leveraged buyout if both rivals are heavily debt financed.

Discussion

Dasgupta and Titman (1996) present a model in which product quality decisions interact with capital structure choice: Firms with high leverage decide to produce low quality products unless they can successfully increase market share and build up reputation for high-quality products. If consumers can observe the product's quality only with a time-lag of one period, firms will compete more fiercely for market shares. A firm's predation potential is increasing in the rival's leverage and is decreasing in the own debt level.

Dasgupta and Titman present a model that captures product market dynamics, because the firms interact twice on the basis of prices in the product market. Moreover, the demand side is explicitly integrated into the profit maximizing of the firms. The capital structure decision, i.e. the level of debt, influences the production decision of both firms during the whole time horizon. To distinguish between high and low quality products is an essential feature of certain product markets (e.g. brand versus no-name products) and is, thus, of high empirical relevance.

However, it is rather unclear why Dasgupta and Titman assume that in the "normal case" firms usually produce high-quality products and switch to low-quality production only as an exemption. In our opinion it would be more convincing to take low-quality production as the reference case and then to let the firms gradually improve the quality of their products.

2.2.2.5 Summary

The approaches of Dasgupta and Titman (1996, 1998) presented above show that the impact of capital structure on dynamic price competition depends on *(i)* the type of uncertainty influencing a firm's profits, *(ii)* on the consumers' preferences for the product variants supplied, *(iii)* on the product quality aimed at and *(iv)* on the debt-equity structure of the competitors. Firms take up debt for strategic reasons when this strategy commits to higher output prices, which will increase the firm's profits. However, a firm will not further increase its debt level if this invites the rival to predatory pricing.

The theoretical results correspond to the empirical findings of Chevalier (1995) and Phillips (1995), who demonstrate that product prices typically increase following a leveraged buyout.

However, they also find evidence that in case one self-financed firm competes with a highly leveraged rival, product prices tend to be low. This phenomenon can be explained by the approach of subsection 2.2.2.4, which states that a self-financed firm will start to aggressively cut its price after having observed the debt structure of its competitor.

In the previous two subsections we have discussed the impact of capital structure on pricing and output strategies in a variety of settings. In each scenario, competition took place between two established firms. In the next subsection, we turn to a different scenario and study market entry. We will investigate the impact of capital structure on product market competition, if there is only one incumbent in the market while the other firm attempts to enter.

2.2.3 Capital structure and market entry

The model of Poitevin (1989) differs in two essential aspects from the previous approaches. First, competition does not take place between two established firm, but between an entrant and a monopolist. Second, Poitevin considers not only external debt, but also external equity financing. Firms can either apply for bank loans or issue new stocks. As we already know, the firms' capital structure decisions affect the competitive strategies in the product market. In addition to this, the capital structure serves as as quality signal to financial investors.

Poitevin shows that a low-cost entrant will always take up debt to signal its quality - although debt increases its bankruptcy probability and makes the firm vulnerable to price wars and predation from part of the incumbent.

2.2.3.1 Assumptions

A1: Players: There is one incumbent firm i in the market. A young firm y threatens this monopoly by its attempt of market entry. If entry is successful, both firms compete first in capacities K_i, K_y, then in prices p_i, p_y. On the financing side, there are rational debt and equity investors.

A2: The game consists of four stages: First, the young firm selects its appropriate debt and equity structure and decides whether to enter the market or not. Second, the incumbent chooses its capital structure. Thus, the financing decisions are characterized by sequential moves. Third, both firms simultaneously choose their capacities. Finally, firms simultaneously determine their optimal output prices.

A3: Information structure: Initially, only the incumbent's marginal cost of capacity is common knowledge, whereas the young firm's cost level is private information. Thus, it is initially unclear which firm has a competitive advantage (see also Milgrom and Roberts 1982). Uninformed players, i.e. the incumbent and potential investors, believe with probability α that the young firm produces with low marginal costs of capacity c_{Ky}^L, and with probability $1-\alpha$ that it produces with high marginal capacity costs, c_{Ky}^H. The young firm's marginal capacity costs can be either higher than or lower than /equal to the incumbent's marginal costs, i.e. $c_{Ky}^H > c_{Ki} \geq c_{Ky}^L$. The young firm's private information is revealed after the second stage, i.e. after both firms have decided upon their financial structure.

Moreover, after capacities are chosen on the third stage, a random variable z positively influences each firm's operating profits as an additive, firm-specific shock. This random variable z ressembles the positive liquidation value in the model of Dasgupta and Titman (1998) above. The random variable leaves the marginal profit of capacities unaffected, but it will influence the bankruptcy region of a firm if this firm has risky debt. The random variables are uniformly distributed over the interval $z_i, z_y \in [\underline{z}; \overline{z}]$, they have the density function $\varphi(z)$ and the distribution function $\Phi(z)$.

A4: All players are risk-neutral.

A5: Production technology: Each firm needs an initial investment I_i, I_y and then produces with constant marginal capacity costs c_{Ki}, c_{Ky}. Under perfect information and in the absence of predation, a low-cost type entrant has a positive expected firm value, while a high-cost type has a negative expected value, if the firms are all equity-financed.

A6: The inverse market demand is $P(K)$, with $K=K_i+K_y$ as the aggregate capacity, $dP(K)/dK<0$ and $d^2P(K)/dK^2 \leq 0$.

A7: Financial policy: Each firm decides upon how to finance its initial investment I_i, I_y. The firms either raise outside equity E in the capital market, where N_N is the number of new shares issued; or borrow D from banks, where the repayment $R=D(1+r)$ includes the interest payment for the amount borrowed. The financial policy is defined by a four-component vector $t_{i,y} = (E,N,D,R)$. Since the financing choice cannot be reversed costlessly within that period, the financial structure decision implies a credible commitment. At the very beginning of the game, both firms hold no debt and have no financial slack.

The model

To highlight the main effects of the capital structure decision, we do not further illustrate the price competition stage (for reference see 2.2.1.2). Instead, we immediately proceed to the capacity stage. Firm i's profits in reduced form are given i as (for firm y analogously)

$$\Pi_i(c_{K_i}, K_i, K_y, z_i) = [P(K) - c_{K_i}] \cdot K_i + z_i \tag{2.29}$$

where K equals total market capacity.

The break-even point from the operating business, i.e. the critical state of nature $\hat{z}_i$ where the firm's operating return just meets the debt obligation, is implicitly defined by the cash-flow equation:

$$-I_i + \Pi_i(c_{K_i}, K_i, K_y, \hat{z}_i) + D + E - R = 0. \tag{2.30}$$

It states that the initial investment I_i must be financed with either debt D or outside equity E or a mixture of both. The firm is solvent in all states higher than $\hat{z}_i$, and must declare bankruptcy if $z_i < \hat{z}_i$.

Moreover, it is assumed that a non-bankrupt firm will obtain an exogenous benefit $B>0$ if its rival goes bankrupt at the end of the period. Since bankruptcy is a lengthy process which retards future production and investment decisions, bankruptcy will give a strategic benefit to the rival. The benefit can also be interpreted as the difference between monopoly and duopoly profits.

The expected equity value of firm i is given by

$$V_i(t_i, t_y, c_{K_i}, c_{K_y}, K_i, K_y) = \left\{ (1 - \Phi(\hat{z}_y))\Phi(\hat{z}_i)B + \int_{\hat{z}_i}^{\bar{z}_i} [\Pi(c_{K_i}, K_i, K_y, z_i) + D + E - I_i - R] d\Phi(z_i) \right\} \tag{2.31}$$

and for firm y respectively. The first term inside the braces represents the expected benefit accruing from a possible bankruptcy of the rival. The second term implies that stockholders are residual claimants in the states in which the firm is solvent.

If the firm issues equity to finance the initial investment, the firm's equity value is divided between old and new shareholders. The expected returns V^0 to the firm's original shareholders are

$$V^0 = \frac{N_0}{N_0 + N_N} \cdot V_i, \tag{2.32}$$

where N_0 represents the number of shares of the original stockholders and N_N the number of shares issued to new stockholders.

The expected returns V^N to the firm's new shareholders are given by

$$V^N = \frac{N_N}{N_N + N_0} \cdot V_i - E. \tag{2.33}$$

New shareholders invest the amount E into the firm and obtain in exchange the complementary fraction of the original shareholders' return. Recall that investment expenditures E of the new shareholders are sunk when the optimal output quantities are chosen.

In case the firm takes up debt, the expected return to the bank is given by:

$$W(t_i, t_y, c_{K_i}, c_{K_y}, K_i, K_y) = -D + (1 - \Phi(\hat{z}_i))R \\ + \int_{\tilde{z}_i}^{\hat{z}_i} [\Pi(c_{K_i}, K_i, K_y, z_i) + D + E - I_i] d\Phi(z_i). \tag{2.34}$$

The bank provides a loan of size D, obtains a fixed repayment R in solvent states and is residual claimant in bankruptcy states.

2.2.3.2 Output stage

The capacity is chosen to maximize the expected value of the firm's equity holders at the time the decision is made. Therefore, the firm maximizes its equity value which accrues to both old and new shareholders, V_i.

$$\max_{K_i} V_i = [1 - \Phi(\hat{z}_i)] \Phi(\hat{z}_y) \cdot B + \int_{\hat{z}_i}^{\bar{z}_i} [\Pi(..) - I_i + D + E - R] \varphi(z_i) dz_i. \tag{2.35}$$

Moreover, Poitevin assumes that each firm can borrow against the potential benefit from the rival's bankruptcy. This causes a shift from the firm's operating break-even point $\hat{z}_i$ to the overall break-even point $\tilde{z}_i$, with $\tilde{z}_i < \hat{z}_i$. Thus, we can rewrite equation (2.35) with the new lower-limit of integration as:

$$\max_{K_i} V_i = \int_{\tilde{z}_i}^{\bar{z}_i} [\Pi(..) + \Phi(\tilde{z}_y) B - I + D + E - R] \varphi(z_i) dz_i. \tag{2.36}$$

The first-order condition for the optimal output strategy is given as follows:

$$\frac{\partial V_i}{\partial K_i} = \left(\frac{\partial P(K)}{\partial K_i} K_i + P(K) - c_{K_i} + B \cdot \varphi(\tilde{z}_y) \frac{d\tilde{z}_y}{dK_i} \right) [1 - \Phi(\tilde{z}_i)] \stackrel{!}{=} 0. \tag{2.37}$$

Here, we have to distinguish between two different cases:

(i) If the rival firm has not issued any risky debt, there will be no bankruptcy benefits to firm i. In this case, the last term in the braces of (2.37) vanishes and the first-order condition reduces to:

$$P'(K) \cdot K_i + P(K) - c_{K_i} = 0 \qquad (2.38a)$$

(ii) If the rival firm has issued risky debt, firm i's first-order condition implies that

$$P'(K) \cdot K_i + P(K) - c_{K_i} + B \cdot \varphi(\tilde{z}_y) \cdot (d\tilde{z}_y / dK_i) = 0 . \qquad (2.38b)$$

The last term of (2.38b) is positive, because an unilateral increase in firm i's capacity shifts the rival's bankruptcy point to the right, $d\tilde{z}_y / dK_i = -P'(K) \cdot K_i > 0$. Hence, for an equilibrium with risky rival debt, the expression $P'(K) \cdot K_i + P(K) - c_{K_i} < 0$ must become negative. This implies, in turn, that firm i installs a higher capacity than it would in the all-equity case.

The last term of (2.38b) represents the predatory effect of the output stage: Upon observing that firm y has a leveraged financial structure, firm i increases its own capacity. This reduces firm y's optimal capacity, lowers its cash flow and makes firm y's default more likely. The predation effect is illustrated in the following figure:

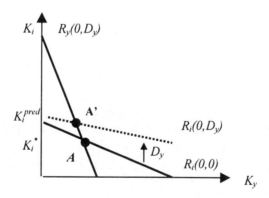

Figure 2.8: Debt-induced predation from part of the incumbent firm
(Poitevin, 1989, 32, and own presentation)

If both firms finance with equity, the simultaneous capacity equilibrium is given at point A. Point A' represents the duopoly equilibrium when firm y, but not firm i is leveraged. Firm i's reaction function is shifted out, thereby raising its own capacity and decreasing its rival's capacity. Predation, thus, reduces the leveraged firm's profit and increases the predator's return.

2.2.3.3 Financial stage

The financial policy is chosen to maximize the old shareholder's equity value, V^0. At the same time, the participation constraints of banks and new shareholders have to be fulfilled. Recall that the young firm's cost type is revealed only after both firms have made their financing decision. Therefore, the low-cost entrepreneur sends a signal to the financial market in order to separate herself from the high-cost type: The low-cost firm issues debt to signal its quality to financiers. The debt level must be high enough to induce the high-cost type to go bankrupt with certainty. As part of the signaling game, the low-cost type may even decide to transfer rents to the debt holders. In Poitevin's model, this is the least-cost separating equilibrium.

By contrast, the incumbent will always choose to issue equity to finance the fixed investment, because he knows that debt entails predation. This is a dominant strategy for the incumbent and represents the best response to any strategy of the young firm.

Proposition 2.7 (Market entry, debt financing, and predation)

The incumbent's dominant strategy is to finance entirely with equity. A low-cost young firm, by contrast, must at least partially finance with debt in order to signal its quality to the financial market, although this quality signal will induce the incumbent to predation.

2.2.3.4 Discussion

Poitevin's (1989) model of capital structure and market entry is very rich and elaborate. Firms choose whether to finance their initial investment with debt or with equity. Both firms have an explicit financing need. Moreover, for each financing alternative, the participation constraints of debt and equity investors are taken into consideration.

As in the previous models, output market equilibrium is affected by the firms' capital structure choice. Poitevin shows that if the young firm issues debt, this will induce an *unleveraged* rival to install a higher capacity. The result stands in contrast to the modified Brander and Lewis (1986) approach, where debt commits the *leveraged* firm to install a higher capacity.

The reason behind is that the young firm does not issue debt because of limited liability and output market considerations, but because an additional type-uncertainty forces the young firm to signal its quality to an input market. Issuing debt, however, has the negative side effect of increasing the young firm's bankruptcy probability. This makes the firm vulnerable to predation from part of the equity-financed incumbent. The incumbent, thus, increases its capacity.

Poitevin provides us with an interesting model of market entry and strategic competition. The framework could easily be adapted to study product innovation and subsequent competition between the young firm and a monopolist.

However, we have serious problems with the financing side: Typically, a young firm does not have access to the stock markets. And if the young firm lacks sufficient track record, firm age, or collateral, it will not even be eligible for bank loan financing. Empirical evidence suggests (see 4.1.1 below), that young firms have to rely on private equity and venture capital instead. In chapter 4, we therefore analyze venture capital financing of a young start-up firm and its impact on product market competition and innovation.

2.3 Empirical findings

So far only a limited number of studies have investigated the interaction between capital structure decisions and product market competition from an empirical point of view.

Opler and Titman (1994) study firms in financial distress and find out that highly leveraged firms loose market shares to their less leveraged competitors. An increase in debt level – especially in R&D intensive industries – leads to an inferior competitive position in the product market.

Phillips (1995) investigates production policy and pricing strategies in markets where the general leverage of firms increased substantially. The increase in debt from 30 percent to 80 percent is generated by leveraged buyout activities. A leveraged buyout is a debt-financed takeover transaction, by which the firm's assets serve as collateral for the required bank loans. Phillips considers four different industries: Polyethylene, fiberglass insulation, tractor trailers, and the gypsum industry. In the first three industries, product prices were rising, while aggregate output declined following the increase in the industry's debt level. This corresponds to the effect of strategic debt on price competition as it was explained by the models of Showalter (1995) and Dasgupta and Titman (1998).

Kovenock and Phillips (1995, 1997) emphasize, too, that the models of price competition fit quite well with the empirical data from the leveraged buyout analyses: According to the theoretical predictions, an increase in debt level will result in higher product prices, and capacities are reduced. This is confirmed by their empirical data for various industries.

Phillips (1995), however, finds one exception in the gypsum industry, which is characterized by capacity competition. In the gypsum industry, market shares of leveraged firms increased, market output increased as well, while prices and profit margins declined. Thus, the phenomena of the gypsum industry can be well explained by the modified Brander and Lewis (1986) model: If firms compete in capacities on a homogeneous market, taking up debt will commit firms to a more aggressive output strategy. The capacity expansion ultimately results in lower market prices and smaller profit margins.

Chevalier (1995) observes the effects of leveraged buyout activities in the supermarket industry. At the end of the 1980s there has been a massive wave of mergers and leveraged buyout transactions within the US-American supermarket industry. Chevalier analyzes how the pricing strategies of supermarkets change following a leveraged buyout. She demonstrates that product prices increase when firms are highly leveraged. These findings correspond to the basic setup of Dasgupta and Titman (1998), who show that product prices rise subsequent to an increase in leverage.

By contrast, Chevalier reports that prices declined if only one of the local supermarkets was highly leveraged, while the competitors were mostly self-financed. Here, the price reduction could be a sign for predatory pricing, which is initiated from the strong, self-financed rivals. This scenario was discussed in the (1996) model of Dasgupta and Titman.

Zingales (1998) finally examines the effect of an increase of debt in the trucking industry. In the mid 1980s the US-American trucking industry underwent an exogenous shock due to deregulation. Deregulation sharply increased the industry's leverage above the desired level, because the value of the firm's operating certificates, a sort of monopoly license, was decreased. As a consequence, the leveraged firms started to charge significantly lower prices than their unleveraged competitors. This result supports the idea that leveraged firms were forced to discount their products, because they were desperate for cash. Thus, the relation between price and leverage depends to a large extent on the nature of the competition and on the financial position of the competing firms.

2.4 Conclusion

In the present chapter we have shown that financial structure matters. In the first part we considered the capital structure decision of an individual firm. In contrast to the Modigliani and Miller proposition we argued that an optimal debt-equity structure exists because of tax benefits of debt interest payments, control contests among stakeholders, agency problems, and asymmetric information between managers and outside investors.

In the second part, we analyzed how a firm's capital structure decision influences the strategic competition in the product market. The key assumption here is that firms which take up debt are protected by limited liability.

In the static framework, we investigated the impact of debt on price competition and on capacity competition. If competition is in prices, firms will only take up debt when demand is uncertain, but not when cost is uncertain. In case of demand uncertainty, strategic debt commits the firms to charge higher product prices, which implies that product market competition will be less intense. If competition is in capacities instead, strategic debt will induce firms to switch to a more aggressive output strategy.

In the dynamic framework, we considered various forms of price competition. In case of price competition and consumer switching costs, firms choose a positive debt level in equilibrium. This induces first-period prices to rise and, in turn, softens the price competition. If, however, product quality is introduced as another strategic variable into the competition game, the results will be reversed: Price competition becomes more intense because firms strategically reduce their first-period prices in order to signal high-quality production and to supply to a large customer base in the second period.

In this scenario, predation becomes possible: If one firm needs external debt financing while the other is self-financed, the self-financed rival may undercut the equilibrium price by just a small margin. The leveraged firm is unable to follow the price cut because of the required debt repayment. As a consequence, the leveraged firm is forced into low quality production.

Leveraged buyout activities provide a rich empirical basis to test the impact of capital structure decisions on product market competition. Highest empirical relevance obtain the models of dynamic price competition.

Nevertheless, we conclude that the interaction between capital structure choice and product market strategy is very sensitive to the type of competition prevailing in the product market.

In the next two chapters, we will investigate the corporate finance decisions of firms more in detail. We leave the simple debt versus equity-choice aside and assume, instead, that firms must apply for external debt or equity contracts on imperfect capital markets. The imperfections of the financial markets are caused by asymmetric information.

The corporate finance decisions are, thus, analyzed in terms of a principal-agent relationship. The entrepreneur as the agent needs external funds to finance an investment project. The entrepreneur signs a financial contract with the principal, i.e. a bank or a venture capital company, who supplies the funds in exchange for a share of the project's return.

The financial contracts will affect the firms' pricing strategies and the equilibrium output in the product market. Chapter 3 is dedicated to the question of debt financing. Chapter 4 addresses the question of private equity, i.e. venture capital financing.

3 Credit financing and strategic competition

> *Whenever one individual depends on the action of another, an agency relationship arises. The individual taking the action is called the agent. The affected party is the principal.*
>
> Pratt and Zeckhauser (1985)

In the present chapter we concentrate on debt financing of innovating and competing firms. Thus, we consider firms that are generally well established in the product market. These firms have sufficient firm age and reputation and are capable to provide collateral for bank loan financing. Debt financing is the most common form of external financing for firms in that stage: Bank loans account for about two thirds of external financing of corporate investments (Deutsche Bundesbank, 2001, 29).

Bank loan financing is a relatively "cheap" form of financial source: The bank obtains only a fixed repayment for its credit extension and does not participate in the firm's upside potential. The bank, though, has to bear the downside risk if the firm's project doesn't succeed. Bank-loan financed investment projects are typically much safer than for example venture capital backed projects, i.e. their default probability is much lower. Credit contracts are standardized financial instruments.

In the models below, we deviate from the neoclassical paradigm of perfect credit markets. Instead, we assume that credit markets are imperfect due to ex-post asymmetric information between borrowers and banks. Thus, we focus on principal-agent relationships in financial contracting and analyze how they affect competition and innovation in the product market.

On the product market side, we focus on dynamic price competition. Moreover we assume, that two well-established firms try to innovate and to realize a process innovation. A process innovation reduces marginal production costs and helps the firm to increase its market share.

The chapter is organized as follows: The Introduction (3.1) briefly describes the main features of contract theory. Section (3.2) investigates the individual firm-bank relationship when financial contracts are influenced by ex-post asymmetric information. We present the basic models of moral hazard, and show how agency problems can be mitigated with long-term debt contracts. A natural question is then to ask, how these capital market imperfections influence the competitive structure in an oligopolistic product market. In section (3.3), we therefore analyze strategic product market competition when firms need external funding and are financed via long-term debt contracts. We present the work of Chevalier and Scharfstein (1996) and Stadler (1997) and demonstrate how capital market and product market imperfections reinforce each other. Neff (1999) finally includes innovation activities into the analysis. Firms compete in prices for market shares. Moreover, they engage in cost-reducing R&D activities in order to survive in competition. The need for bank loan financing alters the competitive strategies in the product market. We show that a financial debt contract causes R&D intensity to decline, while industry concentration tends to rise. Section 3.4 summarizes our results.

3.1 Introduction: Contract theory

Contract theory is a strong analytical tool which we need in the following chapters to analyze the financial relationship between the firm and its investors.

3.1.1 General ideas and definitions

When an external debt or equity investor provides funding for a firm's investment or innovation project, both parties typically sign a contract. In a world without information problems, financial contracts can be made contingent upon all states of nature. If the result $\tilde{\Pi}$ of an investment project is observable by both the entrepreneur and the financial investor at the same time (a situation of symmetric information), they will sign a contract specifying in advance how they share the stochastic return $\tilde{\Pi}$. This sharing rule is completely determined by the repayment $R(\Pi)$ to the financial investor which is expressed as a function of the realization Π of $\tilde{\Pi}$. The contracting parties can, therefore, restrain from monitoring activities or penalties.

However, such a contract is not feasible, if the entrepreneur after observing the profits can appropriate part of the surplus, while only reported profits are the basis of the financial contract. Asymmetric information, thus, makes coordination among contracting parties more difficult.

Agency theory classifies two types of ex-post informational asymmetries:

(i) If the entrepreneur gets access to private information that cannot be altered (e.g. the attained profit level, the quality of a research project, etc.), we speak of moral hazard with hidden knowledge.

(ii) If the entrepreneur has to undertake an action (e.g. to spend effort on the innovation project), which is not observable by the financial investor, a problem of moral hazard with hidden action arises.

The financial contract between the entrepreneur and the financial investor tries to circumvent these moral hazard problems.

Contracts are called "verifiable", if the payoff structure both parties agree upon depends exclusively on parameters which can be verified by courts. In case the parameters or actions are not verifiable by courts, but observable by the other party, the contractual arrangement is characterized by "relationship-specific investments".

A contract is called "complete" if it specifies the payoff structure for all possible states of nature. Technically speaking, a complete contract induces a non-cooperative game, in which rational players look for Nash-equilibria. Given that the information is symmetric at the signing of the contract, the Nash-equilibrium selected by the players will Pareto-dominate all other allocations.

By contrast, a contract is "incomplete" if the payoff structure cannot be specified for all possible contingencies. Since there are many potential states of the world, it may be prohibitively costly to write a contract that conditions on all these states. Contracting parties signing incomplete contracts typically engage in long-term relationships. Due to inconsistencies in interests over time, both sides may want to renegotiate the initial contract in later periods. These renegotiations may restrict the initial contracts to ex-ante agreements which settle only monitoring and enforcement structures, i.e. governance structures (Schweizer 1999, chapter 6). Though incomplete contracts are an interesting and still progressing area of research (Tirole, 1999), the debt and equity contracts of our chapters 3 and 4 are all based on the concept of complete contracts.

3.1.2 The revelation principle

How does the entrepreneur's private information become elicited? How can the optimal contract be found among all possible contractual arrangements? Here, the revelation principle greatly simplifies the analysis:

First, we define the set of possible states (e.g. the possible outcomes of an investment project) within our complete contract framework. Then, in searching for an optimal contract, the financial investor as the principal can restrict himself to contracts of the following form:

3.2 Credit financing: The individual firm-bank relationship

(i) After a certain state has been realized, the entrepreneur is required to announce which state has occurred.

(ii) The contract specifies a payoff scheme for each possible announcement.

(iii) In every state, the entrepreneur finds it optimal to report the state truthfully.

Point *(i)* states that the contract will be a direct mechanism.[1] A direct mechanism simply asks the entrepreneur to reveal her private information. The contracts is, then, based solely upon the entrepreneur's report.

Ad *(ii)*: The announcements may be true or false. Nevertheless, the payoffs are assigned according to the entrepreneur's various announcements. Both contracting parties commit to this mechanism and do not revise it after hearing the report.

Ad *(iii)*: The payoff scheme must be such that the entrepreneur finds telling the truth to be an optimal response. Revelation mechanisms with this truthfulness property are known as incentive-compatible. In the financial contracts to be analyzed below, truthful reporting is guaranteed by the incentive-compatibility constraints.

Thus, the revelation principle tells us that in our search for the optimal contract we can restrict attention to direct revelation mechanisms in which truth telling is an optimal strategy for the agent. Given these assumptions, the optimal financial contract is the solution to the following program:

Maximization of the principal's or the agent's objective function.

(The objective function depends on which party has the bargaining power)

Subject to:

- the participation constraint of the financial investor,
- the incentive-compatibility constraint of the entrepreneur (truthful reporting is required), and
- the wealth constraint of the entrepreneur.

3.2 Credit financing: The individual firm-bank relationship

In this section we analyze how asymmetrically distributed information affects the equilibrium results in the credit market. As credit market we define the segment of financial markets where firms and banks meet to exchange demand and supply of loans. When the credit market is incomplete due to asymmetric information, the interest rate does not play its neoclassical role as market clearing mechanism any

---

[1] A mechanism is defined as a collection of strategy sets and a payoff function.

longer. Instead, the interest rate acts as an incentive mechanism in the sense that a change in the interest rate induces a change in the agent's economic behavior. This may cause a problem of moral hazard.

3.2.1 Moral hazard

> *One should hardly have to tell academics that information is a valuable resource: Knowledge is power.* George Stigler (1986)

In this subsection we analyze the consequences of ex-post asymmetrically distributed information on financial contracting. Townsend (1979) and Gale and Hellwig (1985) were the first to derive debt contracts as the optimal solution to an incentive problem under asymmetric information. Here, only the investing firm, but not the credit extending bank is able to observe the actual amount of profit at no cost, such that a problem of moral hazard arises: The entrepreneur can either falsely report low profits, or may divert money to her private ends before the bank knows which level of profits has been realized. Banks, therefore, engage in monitoring activities in order to verify the outcome of the entrepreneur's project. When costly state verification is required for numerous investment projects, a bank has advantages over individual lenders, because the bank can realize higher economies of scale concerning the monitoring activities. According to Diamond (1984), this justifies the existence of a banking system.

3.2.1.1 Costly state verification and standard debt contract

We present the model of Gale and Hellwig (1985) because of its simplicity and elegance. The main features referring to the standard debt contract can also be found in Townsend (1979) and Williamson (1987).

Assumptions

A1: Players: An entrepreneur has a profitable investment opportunity, but needs external funds to finance it. A competitive bank is willing to supply a loan as long as the expected repayments are equal to $R=(1+r)D$, where D is the face value of debt and r is the market interest rate for debt. The entrepreneur is subject to limited liability.

A2: The time horizon is one period.

A3: Information structure: The project's return is stochastic. Asymmetric information arises because the entrepreneur obtains private information

3.2 Credit financing: The individual firm-bank relationship

about the actual level of profits. The bank may verify the profit realization, which induces fixed monitoring costs of m.

A4: The bank and the entrepreneur are both risk neutral.

A5: Production technology: The project requires a fixed investment I, which is completely debt financed, $I=D$. The project yields a continuously distributed gross return of $\Pi \in [0;\Pi^{max}]$. The repayment obligation R lies between 0 and Π^{max}, so that with positive probability the firm cannot pay back its loan. The project, though, is profitable, i.e. $R \leq E(\Pi)$.

A6: The demand side of the firm's product market is not explicitly considered.

A7: The financial contract has to specify three features: *(i)* the repayment R, i.e. a transfer from the borrower to the lender based on reported profits $\hat{\Pi}$; *(ii)* the monitoring rule $J \in \{0,1\}$, i.e. the bank's decision when to undertake a costly audit; and *(iii)* the repayment to the lender after costly monitoring has taken place. Using the revelation principle (see Fudenberg and Tirole 1991, Mas Colell, Whinston, and Green 1995) the contract can be described as a direct revelation mechanism under which the entrepreneur is asked to report the realized profits. The mechanism has to fulfil the incentive compatibility constraints, ensuring that truthful reporting, $\hat{\Pi} = \Pi$, is a dominant strategy for the entrepreneur. We show that the efficient incentive compatible contract is a standard debt contract.

The model

Gale and Hellwig (1985) define $J_0 = \{\Pi \mid J(\Pi)=0\}$ as the region of profits in which no verification takes place, and $J_1 = \{\Pi \mid J(\Pi)=1\}$ as the region of profits where costly monitoring is undertaken. If monitoring shows that the entrepreneur has made a false report, all the cash flow will be transferred to the bank. Incentive compatibility requires:

In the non-auditing region, i.e. if $\Pi \in J_0$, the repayment function $R(\Pi)$ must be designed in such a way that the entrepreneur truthfully reports her profits. Thus, it must hold that truthful reporting makes the entrepreneur better off than lying:

$$\Pi - R(\Pi) \geq \Pi - R(\hat{\Pi}) \quad \forall \, \Pi, \hat{\Pi} \in J_0, \tag{3.1}$$

where $\hat{\Pi}$ stands for reported profits. This is equivalent to $R(\Pi) \leq R(\hat{\Pi})$ and $R(\hat{\Pi}) \leq R(\Pi)$ for all $\Pi, \hat{\Pi} \in J_0$, which, in turn, implies that in the non-audit region, the entrepreneur's repayment obligation must be a constant.

$$R(\Pi) = R(\hat{\Pi}) = R_0 \quad \forall \, \Pi, \hat{\Pi} \in J_0. \tag{3.2}$$

Intuitively it is clear that the repayment function is necessarily a constant, since otherwise the borrower could cheat by announcing profits that correspond to the minimum repayment in this non-audit region.

In the monitoring region, i.e. if $\Pi \in J_1$, the repayment function $R(\Pi)$ must be specified in such a way that the entrepreneur has no incentive to falsely report that profits are realized in the non-monitoring region:

$$\begin{aligned} \Pi - R(\Pi) &\geq \Pi - R(\hat{\Pi}) \quad \forall \ \Pi \in J_1, \ \hat{\Pi} \in J_0, \\ &= \Pi - R_0 \quad \forall \ \Pi \in J_1. \end{aligned} \quad (3.3)$$

Thus, it must hold:

$$R(\Pi) \leq R_0 \quad \text{with} \quad R(\Pi) = \Pi \quad \forall \ \hat{\Pi} \in J_1. \quad (3.4)$$

The optimal contract, therefore, solve the entrepreneur's maximization problem:

$$\max_{R(\Pi), J(\Pi)} \int_0^{\Pi^{max}} [\Pi - R(\Pi)] dF(\Pi), \quad (3.5)$$

under the investor's participation constraint (recall that m represents the fixed monitoring costs):

$$\int_0^{\Pi^{max}} [R(\Pi) - m \cdot J(\Pi)] dF(\Pi) = (1+r)D, \quad (3.6)$$

under the entrepreneur's incentive compatibility constraints

$$\begin{aligned} R(\Pi) &= R_0 \quad \forall \ \Pi \in J_0 \\ R(\Pi) &= \Pi \quad \forall \ \Pi \in J_1, \end{aligned} \quad (3.7)$$

and under the entrepreneur's wealth constraint

$$\Pi - R(\Pi) \geq 0. \quad (3.8)$$

An efficient incentive-compatible debt contract is obtained by minimizing the probability of monitoring for a fixed expected repayment, or, equivalently, by maximizing the expected repayment for a fixed probability of monitoring. This implies that a costly audit will take place only when the repayment is less that R_0. This can be interpreted as a standard debt contract. We summarize these findings in the following proposition.

Proposition 3.1 (Standard debt contract)

The standard debt contract is an incentive-compatible contract which is characterized by the following features:

1. *When the firm is solvent, the contract involves a fixed repayment of R_0 to the bank. The bank does not participate in investment profits above R_0.*
2. *If the firm cannot meet its repayment obligations, it is declared bankrupt. In this case, the bank engages in costly state verification which induces monitoring costs of m.*
3. *In case of bankruptcy, the firm has to transfer all its profits to the bank.*

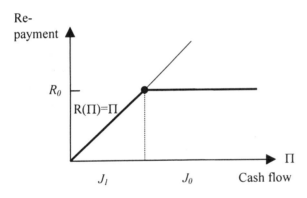

Figure 3.1: The standard debt contract, with J_1 denoting the monitoring region, J_0 the non-monitoring region. Own presentation, see also Freixas and Rochet (1998, 97).

Discussion

The model of Gale and Hellwig (1985) obtains as result the standard debt contract, a strong result, which is based on the existence of verification costs. The verification costs can be formulated in a very general way: They can be fixed as in our presentation, or they could vary with the profit level $m(\Pi)$.

The existence of verification costs, however, leads to an inefficiency: Investment projects with a positive net present value will not be financed, if the expected verification costs exceed the project's value. Moreover, if the costs of state verification are too high, no loan agreement will be reached at all, and in this case it is impossible to solve the information problem.

Although the standard debt contract allows for an incentive-compatible credit relationship between the firm and the bank, it must be criticized for a problem of time inconsistency: From the dynamic, game-theoretic point of view, an automatic verification by the bank in case of bankruptcy is not a credible strategy-profile, because monitoring costs imply a dead-weight loss to the bank. The firm, on the other hand, will never lie because it is faced with the permanent threat of verification in the bankruptcy case. This means that both contracting parties could improve their overall situation, if the bank verified the state not automatically, but stochastically. Thus, the loan contract should incorporate mixed strategies of monitoring activities. We admit, however, that these are difficult to write down in a loan contract.

Another way to solve the time inconsistency problem is to allow the investor to choose between monitoring and debt renegotiation, as it is done in the model of Bester (1994) below.

3.2.1.2 Renegotiation of the standard debt contract

Bester (1994) discusses credit contracts with moral hazard, bankruptcy, and renegotiation. The model builds on the framework of Gale and Hellwig (1985), but differs from it in two ways: *(i)* First of all, the project's return is dichotomous instead of continuously distributed: With probability θ the project is successful and the return realization is Π^{max}, whereas with probability *(1-θ)* the project fails and the return is $\Pi^{min} > 0$. Since return realizations are again not verifiable to outsiders, the borrower's repayment obligation R_0 cannot be conditioned on the project outcome. *(ii)* Second, the game is extended by a renegotiation stage to overcome the inefficiencies arising from the bankruptcy event: In case the entrepreneur reports low profits, the firm will not be declared bankrupt, if the entrepreneur agrees to transfer the minimum profit Π^{min} plus some previously specified collateral C to the bank. It is assumed, however, that the sum of collateral and minimum profit does not cover the total amount of loan extended $\Pi^{min} + C < D < R_0$. Nevertheless, the firm can obtain a partial debt relief, whereas the bank is saving on monitoring costs.

The renegotiation game between firm and bank is formulated in three steps.

1. In stage one of the borrower-lender relationship, the return Π^{max} is realized with probability θ, and Π^{min} is realized with probability *(1-θ)*. This is observed by the entrepreneur, while the bank remains uninformed.

2. In stage two, the entrepreneur decides upon repayments: In case of project failure, the entrepreneur is forced to default. In case of success, she can choose between truthful repayment or strategic default. The entrepreneur chooses a possibly mixed strategy so that she repays R_0 with probability τ and lies with probability *(1-τ)*.

3.2 Credit financing: The individual firm-bank relationship

3. Third, if the repayment obligation is not fulfilled, the bank will decide whether to impose bankruptcy or to renegotiate the contract. Bankruptcy entitles the bank to liquidate the entire project. Renegotiation implies that the bank offers a modified contract $\Gamma'(\Pi^{min}, C)$ to the firm, which reduces the firm's debt obligation to Π^{min} and transfers the collateral C to the bank. Again, we allow for random strategies, where $(1-\eta)$ denotes the probability for renegotiation, while η denotes the probability of bankruptcy.

Figure 3.2 illustrates the players' moves in the renegotiation game and describes the respective payoffs to both the firm and the bank.

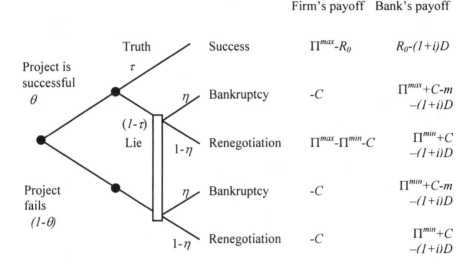

Figure 3.2: Moral hazard and renegotiation of the debt contract.
Own presentation, adapted from Bester (1994, 77)

If the firm correctly repays its loan, then ex-post there is full information because the project must have been successful. Moreover, if a perfect Bayesian equilibrium exists, the bank's and the firm's expectations must be consistent. An equilibrium in mixed strategies, thus, implies, that the bank is indifferent between monitoring (η) and subsequent liquidation of the firm on the one side, and renegotiation $(1-\eta)$ and partial debt relief on the other side. In addition to this, the firm is indifferent between cheating $(1-\tau)$ and reporting profits truthfully (τ). In case the firm reports low profits, costly state verification is not automatically undertaken, but takes place only with a certain probability. The bank avoids monitoring costs, and the firm receives higher expected profits because of the reduced bankruptcy probability.

Note that the role of collateral here is different from its function in Bester (1985): In the earlier approach, collateral served as a screening device to sort loan applicants into different risk categories. Empirical evidence suggests, however, that collateral is more likely to be used for financing high-risk investments. In the present model, the option to renegotiate the initial contract seriously undermines the role of collateral as a screening device: The prospect of debt renegotiation no longer induces truth-telling behavior on the part of the entrepreneur. Collateral does not punish for project failure, but instead makes default less attractive in the event of success.

3.2.2 Multi-period credit contracting

The extension to a multi-period framework generally reduces the problems of asymmetric information: With repeated bank–firm relationships, the firm has an incentive to acquire the reputation of being a reliable debtor and to avoid cheating.

3.2.2.1 Dynamic moral hazard

Chang (1990) studies a two-period extension of the costly state verification model of Gale and Hellwig (1985, see subsection 3.2.2.1 above). An entrepreneur wants to invest the amount I into a project which produces two independent random cash flows, Π_t, $t=1,2$. The entrepreneur has no initial wealth and, therefore, applies for a bank loan. At the end of each period, the entrepreneur is privately informed about the project's return. The bank observes the cash flow realizations only at an auditing cost, $m_t(\Pi)$, $t=1,2$. If the entrepreneur retains some earnings at $t=1$, and the bank monitors at $t=2$, the bank will discover any potential diversion of funds. The question is to determine the optimal debt contract that minimizes the expected monitoring costs of an audit for a given expected repayment to the lender. Use of the revelation principle implies that the contract depends directly on the entrepreneur's announcements $\hat{\Pi}_t$, $t=1,2$. The contract specifies $\{R_1(\hat{\Pi}_1); J_1(\Pi_1); R_2(\hat{\Pi}_1, \hat{\Pi}_2); J_2(\Pi_1, \Pi_2)\}$, where R_t indicates the entrepreneur's repayments to the bank, and J_t represents the dichotomous variable for monitoring: $J_t=1$ if an audit occurs at date t, and $J_t=0$ otherwise.

Since the time horizon ends after the second period, backward induction implies that the second-period contract is necessarily a standard debt contract, which specifies a repayment R_2 that is independent of Π_2. An audit will take place at date 2 if and only if the total cash flow of the firm $\Pi_1 - R_1(\hat{\Pi}_1) + \Pi_2$ is less than R_2, in which case the bank seizes all the cash flow. If monitoring costs were constant, it would never pay to monitor the firm in the first period, since the firm's cash flow cannot be diverted away. The problem will be interesting only if the verification cost m_2 is strictly increasing with the firm's sum of cash flow. Then, it is never optimal to let the firm retain some of its earnings at the first period, because this would increase the monitoring costs in the second period.

Chang shows that the optimal contract defines three regions: *(i)* If the first-period cash flow is below a certain constant, i.e. $\Pi_1 \leq l$, monitoring takes place, the repayment equals $R_1 = \Pi_1$, and the entrepreneur is required to pay back the remaining debt, $R_2 = R^{total} - \Pi_1$, in the second period if she can. *(ii)* If $l < \Pi_1 < R^{total}$, there is no monitoring, but the entrepreneur repays as much as possible after the first period, i.e. $R_1 = \Pi_1$, and in the second period she repays $R_2 = R^{total} - \Pi_1$. *(iii)* If $\Pi_1 \geq R^{total}$, all debt is repaid after the first period. Thus, the optimal dynamic contract can be interpreted as a two-period standard debt contact with an early repayment option after the first period.

Similarly, Webb (1992) investigates two-period financial contracts with private information and costly state verification. He shows that under a long-term contract the entrepreneur is induced to report more honestly that than under a sequence of short-term contracts.

3.2.2.2 Long-term contracting when costly state verification is not possible

Bolton and Scharfstein (1990) study a repeated borrower-lender relationship in which the termination threat from part of the bank provides an incentive for the entrepreneur to repay the loan. Bolton and Scharfstein consider a two-period model of moral hazard where - in contrast to the previous analysis - costly state verification is not possible. The model is very important because this form of debt contract will be used in section (3.3) to study interaction between capital market imperfections and product market competition.

Assumptions

A1: An entrepreneur has a two-period investment project, which requires an investment I_t in each period $t=1,2$. The entrepreneur has no initial wealth. She therefore applies for a bank loan. The banking sector is competitive. The bank makes a take-it-or-leave-it contract offer to the firm at the beginning of the game. The entrepreneur accepts the contract as long as its expected value is nonnegative. The entrepreneur is subject to limited liability.

A2: The time horizon is two periods. The discount rate is zero.

A3: Information structure: It is common knowledge to both the firm and the bank that production in each period is a positive net present value investment. Gross profits Π_t in each period are observable, but not verifiable by third parties. An equivalent assumption is that the entrepreneur privately observes profits and, thus, has a chance to divert funds for private consumption. The contract, in either case, cannot be made directly contingent on profits.

A4: The bank and the firm are both risk neutral.

3 Credit financing and strategic competition

A5: Production technology: At the beginning of each period, the firm invests the amount I_t in order to start production. The gross profit in each period is either high, Π_t^H, with probability θ, or low, Π_t^L, with probability $1-\theta$. The expected return of the investment in each period is positive, $\theta \Pi_t^H + (1-\theta)\Pi_t^L > I_t$. With positive probability, though, the project looses money, $\Pi_t^L < I_t$. Profits are independently distributed across periods.

A6: Consumer demand is not explicitly given.

A7: Financial contract: All the bargaining power is on the side of the bank, such that all the rents from the financial relationship accrue to the bank. The contract $\Gamma(R_1, \beta, R_2)$ is a direct revelation mechanism which specifies the firm's repayments R_t as well as the refinancing probability for the second period, $\beta \in [0;1]$. The bank can force the entrepreneur to pay out a minimum profit in each period, Π_t^L. The project is stopped if first-period profits are low. This termination threat is the essential part of the contract.

The model

In a one-period model, the bank would not invest into the project, because without costly state verification the firm would always report low profits and its repayment $\Pi_t^L < I_t$ would be insufficient to cover the credit outstanding.

If instead the financial relationship covers two periods, the bank will decide whether the entrepreneur receives funding in the second period. The bank will threaten to cut off financing in the second period, if the entrepreneur defaults in the first. This induces the entrepreneur to repay more than the minimum Π_1^L in the first period. This threat is credible since no other bank wishes to finance the firm in the second period.

Formally, the contract is designed as a direct revelation mechanism in which the terms of contract are based on the firm's reported profits. The optimal contract maximizes the expected profits of the bank subject to: *(i)* The incentive constraint, which states that the firm truthfully reveals its profits in both period $t=1,2$. *(ii)* The limited liability constraints which state that the entrepreneur cannot repay more than the project's returns. *(iii)* The individual rationality constraint, which ensures that the entrepreneur is willing to participate in the contract. The bank's maximization problem is as follows:

$$\max_{R_1, \beta, R_2} W = -I_1 + \theta[R_1^H + \beta^H(R_2 - I_2)] + (1-\theta)[R_1^L + \beta^L(R_2 - I_2)], \quad (3.9)$$

subject to the entrepreneur's incentive constraint

3.2 Credit financing: The individual firm-bank relationship

$$\Pi_1^H - R_1^H + \beta^H [E(\Pi_2) - R_2] \geq \Pi_1^H - R_1^L + \beta^L [E(\Pi_2) - R_2]. \quad (3.10)$$

Equation (3.10) ensures that the entrepreneur truthfully reveals a high project outcome, because the net first and second period profits under a high refinancing probability surpass the profits when she lies and reports low first-period profits instead. We omit the second incentive constraint under which the entrepreneur reports high instead of low profits, because this restriction is not binding.

The static and the intertemporal limited liability constraints require that

$$\Pi_1^H \geq R_1^H ; \quad \Pi_1^L \geq R_1^L ; \quad \Pi_1^H - R_1^H + \Pi_1^L \geq R_2 ; \quad \Pi_1^L - R_1^L + \Pi_1^L \geq R_2. \quad (3.11)$$

Lastly, the individual rationality constraint states that the entrepreneur's expected net profit from the two-period investment should be positive:

$$\theta(\Pi_1^H - R_1^H + \beta^H [E(\Pi_2) - R_2]+ \\ (1-\theta)(\Pi_1^L - R_1^L + \beta^L [E(\Pi_2) - R_2] \geq 0. \quad (3.12)$$

The maximization problems is solved by backward induction. From the argument presented above it is clear that the second-period repayments cannot depend on second-period profits. The optimal contract thus specifies that second-period repayments in the high and low case will be identical and amount to $R_2 = \Pi^L$. The bank can at most extract Π^L from the firm because there is no termination threat at the end of the second period. Moreover, the incentive constraint (3.10) is binding in the optimum due to the distribution of bargaining power. These two results simplify the bank's maximization problem to

$$\max_{R_1, \beta} W = -I_1 + R_1^L + \beta^H \theta [E(\Pi_2) - I_2] - \beta^L [\theta E(\Pi_2) + (1-\theta)I_2 - \Pi_1^L]. \quad (3.13)$$

It follows immediately that refinancing after a high project outcome should be as high as possible, i.e. $\beta^H = 1$. On the other hand, the last bracketed term is positive since $E(\Pi_2)$ and I_2 both exceed Π_1^L. Therefore, refinancing after a low project outcome should be as low as possible, i.e. $\beta^L = 0$. From the limited liability constraint we derive that $R_1^L = \Pi_1^L$. Then, it follows from the incentive compatible constraint that $R_1^H = E(\Pi)$. Finally, we must determine the conditions under which the bank earns nonnegative profits. Given the optimal contract, the bank's expected profits are $-I_1 + \theta E(\Pi) + \Pi^L - \theta I_2$. Thus, for the bank to grant a credit at the beginning of the game, the investment I_t can be no greater than

$$I_t \leq \frac{\theta E(\Pi) + \Pi^L}{1+\theta}. \quad (3.14)$$

As a consequence, some positive net present value projects may not be funded. We summarize our results in the following proposition.

Proposition 3.2 (Long-term contract when costly monitoring is not possible)

The bank will offer a long-term financial contract if and only if the investment in each period does not exceed $I_t \leq [\theta E(\Pi) + \Pi^L]/(1+\theta)$. In this case, the optimal financial contract specifies $\{R_1^L = \Pi^L, \beta^L = 0, R_1^H = E(\Pi), \beta^H = 1, R_2 = \Pi^L\}$. The firm will operate in the second period if and only if first-period profits are high.

The proposition shows that there is an ex-post inefficiency because the firm is liquidated when first-period profits are low, although expected profits from the second period are positive, $E(\Pi) > I_2$. Thus, after a first-period profit of Π_1^L has been realized, the entrepreneur would naturally wish to renegotiate the contract. Note, however, that even though it is efficient to produce, the bank can at most receive $\Pi^L < I_2$ due to the information problems between borrower and lender. Thus, it is impossible to renegotiate the initial contract;[2] and no other bank is willing to provide the money for the second-period investment.

Discussion

The assumption that costly state verification is not possible is quite strong. Financial auditing should, at least for bigger companies, be no problem, because the firm is legally obliged to establish a valid accounting system. Bolton and Scharfstein (1990), however, suggest that irregular accounting practices can make it difficult for outside investors to know the firm's true profitability. Moreover, it is often difficult to judge whether particular expenses are necessary or not. What looks like justifiable expenses may really be managerial perquisites with no productive value. Especially if we interpret the firm's project as a novel research and development project, innovation costs and the entrepreneur's personal expenditures may not always be very easy to separate.

A minor remark considers the refinancing probability β. In order to guarantee a credible termination threat there must exist an enforceable randomizing scheme allowing β-values to lie between zero and one.

---

[2] Renegotiation undermines the incentive compatibility constraint. If rational parties anticipate that they can renegotiate the contractual terms after the first period, an additional constraint is imposed to the bank's maximization problem, which, in turn, further reduces market efficiency. Thus, a contract should be designed in a way that renegotiation becomes difficult. One solution here is to enhance the number of creditors, a scenario investigated by Bolton and Scharfstein (1996).

3.2 Credit financing: The individual firm-bank relationship

As mentioned above, when first-period profits are low, the firm will be liquidated after the first period, which is inefficient. Liquidation, however, is necessary to induce the entrepreneur to truthfully report high profits and to repay the loan. The reason behind this is that the bank cannot distinguish between liquidity default, i.e. when low profits are due to an unfavorable state of nature, and strategic default, i.e. when the entrepreneur just pretends that profits are low. Thus, the bank must also deny refinancing in case of liquidity default.

Note that the contract is signed under symmetric information. The inefficiency results from the entrepreneur's ex-post chance to divert funds at the expense of the bank, as well as from the entrepreneur's limited liability and wealth constraints.

3.2.3 Discussion

In section (3.2) we have seen that asymmetrically distributed information between borrower and lender requires a complex design of the loan contracts. The assumptions which underly the moral hazard models are quite plausible: Banks do, typically, not get involved in the day-to-day business of a firm. Thus, the firm can costlessly divert some of the operating cash flow. Managers can spend corporate resources on pet projects or empire building. Such spending cannot be directly controlled through contractual means. Here, a long-term relationship may help to mitigate the moral hazard problem. If the firm and the bank repeatedly interact, the entrepreneur may wish to build up a reputation for being a reliable debtor. This may serve to obtain refinancing for another investment project (see Bolton and Scharfstein (1990), or the firm will obtain better terms of credit if it proves to be a reliable debtor and doesn't cheat on the financial contract (see Maksimovic 1990).

So far, we have focused on the bilateral relationship between a firm and a bank. In economic reality, firms are not isolated units, but stand in interaction with other firms of their industry. In the next section we therefore analyze the impact of capital market imperfections on a firm's competitive position in the product market.

3.3 Credit financing and product market competition

> *Information and incentive problems in the capital market can determine the structure of the product market.*
> Bolton and Scharfstein (1990, 94)

On the industry level, we now investigate how the access to financial resources will influence product market competition, investment decisions and innovation activities of firms.[3]

The notion of investment shall here be understood in a broad sense. In his empirical article on "Capital market imperfections and investment", Hubbard (1998) describes the influence of capital market imperfections on such different activities as investment in pricing strategies, investment in research and development, investment in business formation and survival in competition, investment in inventory and investment into human capital of the working force.

In the present analysis, we concentrate on investments in price reduction, in the acquisition of market shares, and in cost-reducing innovation.

Firms in need of external funding are financed with debt contracts in the design of Bolton and Scharfstein (1990). This dynamic approach is necessary, because we assume that costly state verification of the firm's profits is not possible. As we know from the previous section, problems of moral hazard can be mitigated with a long-term contract. Moreover, long-term credit-contracts are required by our industrial organization models of repeated product market interaction.

Section (3.3) is organized as follows: We first present the models of Chevalier and Scharfstein (1996) and Stadler (1997), who investigate the impact of debt contracts if firms invest in market shares. Then, we present our own work based on Neff (1999). In our model, firms compete in prices for market shares. In addition to this, firms engage in R&D activities in order to realize a cost-advantage over their competitor. We show how the financial contracts interact with the firms' pricing strategies and the R&D intensity in that industry.

---

[3] The interaction between financial and product markets is one of the main research topics of the DFG long-term research program 1996-2001 on "Industrial structure and input markets". The central question is how financial decisions of firms influence their investment behavior as well as the competition in the product market. The aim of the theoretical part of the research is to integrate financial (and labor) markets into models of industrial organization.

3.3.1 Credit financing and price competition

In this subsection we investigate the impact of long-term credit contracts on the pricing strategies of firms. Product market competition lasts two periods. Consumers are assumed to be loyal to the firm of their choice.

3.3.1.1 Credit financing, price competition, and consumer switching cost

> *Information and incentive problems in the capital market can limit the ability of cash-constrained firms to make valuable investments.*
> Chevalier and Scharfstein (1996, 704)

Chevalier and Scharfstein (1996) present a two-period model of competition where firms invest in market shares by cutting first-period product prices. Chevalier and Scharfstein show that if firms need external debt contracts to finance their initial production costs, product prices tend to raise.

Assumptions

A1: Players: Two firms i, j compete in prices on a heterogeneous product market. If a firm does not have sufficient internal funds to finance the initial production cost I, it offers a two-period debt contract to a bank. The banking sector is competitive. The two banks who accept the financial contracts are operating independently. The bargaining power is on the side of the firms.

A2: The time horizon is two periods, i.e. firms strategically interact twice in the product market. The discount rate for the second period is zero.

A3: Information structure: Firms face demand uncertainty in the product market. For each firm, the first-period demand shock will be high, z^H, with probability θ, or low, z^L, with probability $(1-\theta)$. While the value of θ is the same for both firms, the actual realizations of the demand parameter z_i, z_j are firm-specific. Expected demand in the first period is $\bar{z}$, while expected demand in the second period is normalized to 1. An informational asymmetry arises between the firm and its bank, because the bank is unable to observe the demand shock nor the firm's profits. In each period, the entrepreneur can costlessly divert all of the cash flow. The bank has no monitoring technology. Thus, the financial contract cannot be made contingent on profit realizations.

A4: Both firms and both banks are risk neutral.

A5: Production technology: In the first period, an initial investment of I is required to start production. Then, each firm produces with constant and identical marginal costs c_t, $t=1,2$. Firms choose their optimal prices before learning the realization of the demand parameter z_i, z_j. Profits are collected after the first and second period.

A6: Consumer demand is formalized according to the Hotelling model (cf. Figure 2.7) in combination with Klemperer's (1995) consumer switching costs approach (cf. subsection 2.2.2.3): Consumers have exogenous tastes for the product variants supplied. They are distributed with uniform density on the line segment [0, 1], with firm i located at 0 and firm j located at 1. Consumers bear a transportation cost T per unit of distance traveled along the line to the product of their choice. This transportation cost, however, must be born only in the first period, in the second period, T equals zero. In the second period, consumers incur a switching cost when buying from the other firm. If this switching cost is high enough, each firm can charge the consumer's reservation price p^r without fear of being undercut by its rival. As long as the price doesn't exceed p^r, each firm's market share in the second period σ_2 is identical to the market share attained in the first period σ_1.

A7: Financial contracting: The financial contract differs from the two-period Bolton and Scharfstein (1990)-contract (see 3.2.3.3) in the following ways: First, in the present paper it is not the threat of reducing the refinance probability, but it is the bank's threat to liquidate the firm after the second period, which induces the entrepreneur to report her profits truthfully. Liquidation, however, is inefficient, because the firm's assets are worth only a fraction $w<1$ in the hands of the bank. Second, the firm needs external debt financing only for the first, but not for the second period of competition.

The model

We start by analyzing the price competition game in the product market, before we turn to the issue of financial contracting. We solve by backward induction.

From the consumer switching cost approach (A5) it follows that second-period profits for each firm i, j depend on their first-period market shares. In particular, we can write firm i's second-period profits when charging the reservation price p^r as

$$\Pi_i^2(\sigma_i^1) = (p^r - c)\sigma_i^1, \tag{3.15}$$

where the first-period market share equals (cf. subsection 2.2.2.3)

$$\sigma_i^1 = \frac{1}{2} + \frac{p_j^1 - p_i^1}{2T} = 1 - \sigma_j^1. \tag{3.16}$$

3.3 Credit financing and product market competition

To calculate the first-period profits, we have to take the expected firm-specific demand shock $\bar{z}_i = \theta\, z_i^H + (1-\theta) z_i^L$ into account:

$$\Pi_i^1(p_i^1, p_j^1, \bar{q}_i) = (p_i^1 - c)\bar{z}_i \sigma_i^1 = (p_i^1 - c)\bar{z}_i \left[\frac{1}{2} + \frac{p_j^1 - p_i^1}{2T} \right]. \qquad (3.17)$$

Firm i chooses the first-period price, given its conjecture about p_j^1, while knowing that the first-period price determines the base of loyal customers who are buying its product in the second period. Firm i, therefore, maximizes over the total profits which amount to

$$(p_i^1 - c)\bar{z}_i \sigma_i^1 + (p^r - c)\sigma_i^1. \qquad (3.18)$$

The first-order condition for this problem defines firm i's pricing reaction curve as a function of firm j's price:

$$p_i^1 = \frac{T + c + p_j^1}{2} - \frac{p^r - c}{2\bar{z}_i}, \qquad (3.19)$$

and analogously for firm j. As is standard, firm i's optimal price is increasing in its rival's price. The symmetric equilibrium price when both firms are self-financed is

$$p^{1*} = T + c - \frac{p^r - c}{\bar{z}_i}. \qquad (3.20)$$

The negative third term of (3.20) demonstrates that the first-period prices are strategically low because firms have to invest into the acquisition of market shares. In a one-period model each firm would charge a price of $T+c$. Here prices are less than $T+c$, because firms compete for first-period market shares so that they can later charge the quasi-monopoly price p^r to their locked-in customers.

Financial contracting between firm and bank

Before starting production, firms need to make an initial investment of I. We assume that firms do not have sufficient internal funds and thus apply for a bank loan of size I. The financial contract calls for a repayment of R^H after the first period; if no such payment is made, the bank has the right to liquidate the firm's assets after the second period (for a similar design of contract see also Bolton and Scharfstein (1996)). The firm will cover its repayment obligation only if the first-period demand shock is high, $\Pi^1(z^H) \geq R^H$. If the first-period demand shock is low, profits are insufficient to repay the loan, $\Pi^1(z^L) \leq R^H$. In this case, the entrepreneur keeps the first-period profits to herself and repays nothing. As a consequence, the firm gets liquidated by the bank and the entrepreneur receives no cash in the second period. The liquidation value of the remaining cash flow is $w \cdot \Pi_i^2(\sigma_i^1)$ with $w<1$. If, by contrast, debt is successfully repaid, the bank has no further leverage over the entrepreneur and the entrepreneur can, therefore, consume all the second-period profits by herself.

The entrepreneur's expected payoff when the initial investment is financed via a loan contract equals

$$V_i = \theta[\Pi_i^1(z_i^H) - R^H + \Pi_i^2(\sigma_i^1)] + (1-\theta)\Pi_i^1(z_i^L). \qquad (3.21)$$

The bank is willing to lend if its expected payoffs are nonnegative:

$$W_i = \theta R^H + (1-\theta)w\Pi_i^2(\sigma_i^1) - I \geq 0. \qquad (3.22)$$

Competition in the banking sector ensures that the participation constraint (3.22) is met with equality.

After the financial contract is signed, firm i chooses p_i^1 taking the repayment obligation as well as p_j^1 as given (firm j analogously). The first-order condition for p_i^1 is

$$\begin{aligned}\frac{\partial V_i}{\partial p_i^1} &= \theta\left[\frac{\partial \Pi_i^1(z_i^H)}{\partial p_i^1} + \frac{\partial \Pi_i^2(\sigma_i^1)}{\partial p_i^1}\right] + (1-\theta)\frac{\partial \Pi_i^1(z_i^L)}{\partial p_i^1} \\ &= \bar{z}_i\left[\frac{1}{2} + \frac{p_j^1}{2T} + \frac{c}{2T} - \frac{p_i^1}{T}\right] - \theta\frac{p^r - c}{2T} \stackrel{!}{=} 0.\end{aligned} \qquad (3.23)$$

With external financing, the entrepreneur only obtains second-period profits with probability θ, while the firm is liquidated with probability $1-\theta$. Thus, a marginal increase in p_i^1 reduces second-period profits by $\theta(p^r-c)/2T$. This means that for each p_j^1, firm i will charge a higher price when it is debt-financed (d.f.) than when it is self-financed (s.f.). The equilibrium price when both firms are debt-financed are then given by

$$p_{d.f.}^{1*} = T + c - \frac{\theta}{\bar{z}_i}(p^r - c) \qquad (3.24)$$

Equations (3.19) and (3.24) imply that for all $\theta<1$ the equilibrium price is higher when firms are debt-financed than when they are self-financed. This implies that with debt-financing, firms are less interested in building market share and they are more short-term oriented.

3.3 Credit financing and product market competition

If only one of the firms is bank-loan financed, while the other is self-financed, the equilibrium price of the bank-financed firm amounts to

$$p^{1*}_{d.f./s.f.} = T + c - \frac{(p^r - c)}{\overline{z}_i} + \tfrac{2}{3}(1-\theta)\frac{(p^r - c)}{\overline{z}_i}; \qquad (3.25)$$

while the first-period equilibrium price of the self-financed firm is lower and equals

$$p^{1*}_{s.f./d.f.} = T + c - \frac{(p^r - c)}{\overline{z}_i} + \tfrac{1}{3}(1-\theta)\frac{(p^r - c)}{\overline{z}_i}. \qquad (3.26)$$

These results are shown in Figure 3.3 below: If a firm is financed via a two-period debt contract, its price reaction function will shift outward.

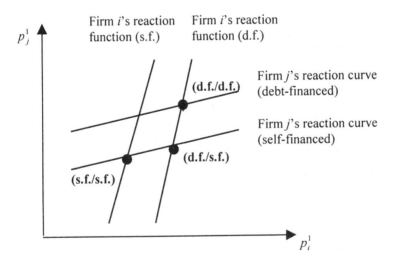

Figure 3.3: Price reaction curves for firms i, j with self-financing and external debt-financing. Own presentation, adapted from Chevalier and Scharfstein (1996, 709. Note that in the original figure there is a mistake in the inscription of the reaction curves).

Proposition 3.3 (Long-term debt contract and investment in market shares)

If a firm is financed by an external debt-contract and firm-specific demand is uncertain, it will invest less in the acquisition of market shares. The first-period equilibrium prices are, therefore, higher than under self-financing.

Debt-financed firms do not reduce their first-period prices as much as self-financed firms in order to acquire market shares. The reason for this is that the credit-financed firm will be liquidated with a certain probability at the end of period one and cannot enjoy the high return of the second period of competition. A debt-financed firm is, therefore, less interested in gaining market shares and, consequently, will invest less in first-period price reduction.

Discussion

The above results correspond to the findings of Dasgupta and Titman (1998) (see 2.2.2.1-2). While in Dasgupta and Titman it was the capital structure, i.e. the firm's debt-equity ratio and the limited liability effect, which lead to an outward shift of the price reaction curves, here it is the two-period debt contract and the asymmetric information between borrower and lender which determines the pricing strategies in the product market. In both cases, external debt financing implies that product prices are higher in the first period and that firms invest less in price reduction. Chevalier and Scharfstein (1996) have, thus, successfully integrated the theory of financial contracting on imperfect capital markets into an industrial organization model of strategic competition on the product markets.

Nevertheless, we have some problems with the financial contract presented above. First of all, it resembles more a static than a two-period debt contract. The firm needs external financing only once, and repayments will be made only once, if the project is successful. In case of failure, the firm gets liquidated in the second period. From a game-theoretic point of view, there is no reason why the bank should wait a period before liquidating the firm.

Second, it seems rather implausible that in bad states of nature, i.e. when the firm fails to meet its repayment obligation after the first period, the entrepreneur will keep $\Pi_i^1(z_i^L)$ all to herself and repay nothing to the bank. Why should she not use the low profits to at least partially repay her debt? We find this assumption too strong to account for general empirical evidence.

A third remark concerns the competition in the product market. Although the consumer switching cost approach is an interesting one, the model is rather static than dynamic, because consumers are locked in in the second period and market shares stay constant. There is no real strategic interaction of firms during the second period of competition.

3.3.1.2 Credit financing, market power of banks, and price competition

> *[B]anks' market power softens the price competition of firms.*
> Manfred Stadler (1997, 376)

Stadler (1997) takes some of the critics mentioned above into account. He investigates the impact of two-period financial contracts on product market competition. Product market competition is formalized in the same way as in Chevalier and Scharfstein (1996). The financial contract, however, is a true two-period contract in the design of Bolton and Scharfstein (1990) (cf. 3.2.3.3). In addition to this, Stadler investigates how market power in the banking sector influences the availability of financial funds, which, in turn, will determine the firms' competitive position in the product market.

Assumptions

A1: Players: Two firms compete in prices with each other on the product market. Two independent, representative banks provide financing if they accept the terms of contract proposed by the firm. The bargaining power is on the side of the firm, which stands in contrast to the original setting of Bolton and Scharfstein (1990).

A2: The time horizon is two periods. There is no discounting.

A3: Information structure: There is uncertainty about the demand in the product market. In each period $t=1,2$, a demand-shock can be high, z^H with probability θ, or low, z^L with probability $1-\theta$. The demand shock ("boom" or "bust") is identical for each firm and independently distributed across periods. The firm privately observes the realization of z and hence has private information about its profits. In case of external debt financing, this induces a moral hazard problem between the bank and the firm.

A4: All players are risk neutral.

A5: Production technology: At the beginning of each period, firms have to make a fixed investment I_t, $t=1,2$. In each period, firms produce with constant marginal costs, $c_i^t = c_j^t = c$. Firms set optimal prices for their heterogeneous products. In the second period, firms charge the reservation price, because consumers are loyal to the product variant chosen in the previous period. Profits are realized at the end of each period.

A6: Consumer demand is formalized according to the two-period model of price competition with consumer switching costs (see Klemperer 1987). In the first period, consumers bear a cost of T per unit of distance traveled along the

Hotelling-line to the product of their choice. This cost measures how far each firm's product variant is from the consumer's ideal set of product characteristics. In the second period, this cost is negligible. Instead, consumers incur a high switching cost if they decide to buy from the other firm. This enables the firms to charge a high reservation price p^r without loosing their customer base.

A6: Financial contracting: If firms need external debt to finance their fixed production costs I_1 and I_2, they will offer a two-period take-it-or-leave-it contract to the banking sector. A bank will accept the contract if expected repayments from both periods are equal to the bank's required minimum profits $\overline{W}$. These minimum profits reflect the degree of market power in the banking sector. Since the bank cannot observe which profits have been realized in each period, the contract must incorporate an incentive-compatible mechanism which ensures that truthful reporting is a dominant strategy for the entrepreneur. This information problem is mitigated by the firm's need to obtain financing for both periods: If the firm announces low profits, $\Pi_1^L(z^L) < I_1$, and cannot meet its first-period repayment obligation, the bank will have the right to reduce or deny follow-up financing in the second period. This threat to stop funding helps to realign the incentives of the entrepreneur. The financial contract is a direct revelation mechanism where repayments are based on the entrepreneur's report of profits. The optimal contract specifies the repayments in case of high and low profits as well as the corresponding refinancing probabilities, $\Gamma(R_1^L, R_1^H, \beta^L, \beta^H, R_2^L, R_2^H)$.

The game is solved by backward induction. That is, we first present the firms' price competition game in the product market, before we derive the optimal financial contract between firm and bank which is signed before production starts.

Product market competition

In each competition period, firms simultaneously choose prices before they learn the current state of demand. From the consumer switching cost approach (see the analysis in the previous subsection) it follows that second-period profits for each firm depend on their first-period market share and on the realization of the demand parameter z. Thus, the respective profits in case of a high or low second-period demand shock are given as:

$$\Pi_2^H(z^H) = z^H(p^r - c)\sigma^1 = z^H(p^r - c)\left[\frac{1}{2} + \frac{p_j^1 - p_i^1}{2T}\right],$$

$$\Pi_2^L(z^L) = z^L(p^r - c)\sigma^1 = z^L(p^r - c)\left[\frac{1}{2} + \frac{p_j^1 - p_i^1}{2T}\right].$$

(3.27)

3.3 Credit financing and product market competition

Correspondingly, high and low profits in the first period are given as:

$$\Pi_1^H(z^H) = z^H(p_i^1 - c)\sigma_i^1 = z^H(p_i^1 - c)\left[\frac{1}{2} + \frac{p_j^1 - p_i^1}{2T}\right],$$

$$\Pi_1^L(z^L) = z^L(p_i^1 - c)\sigma_i^1 = z^L(p_i^1 - c)\left[\frac{1}{2} + \frac{p_j^1 - p_i^1}{2T}\right].$$

(3.28)

Recall that the probability to realize high profits is θ and the probability for low profits is $(1-\theta)$ in each period $t=1,2$.

In case the firms have sufficient internal funds to finance their fixed production costs, no information problem arises and the firm is refinanced with certainty. As usual we assume that expected profits exceed investment costs I_t in each period. Taking total expected profits into account, we derive the first-order conditions for the price reaction functions and solve for the equilibrium prices. The equilibrium prices under self-financing amount to:

$$p^1* = T + c - (p^r - c).$$

(3.29)

We now investigate how external debt financing affects the firm's equilibrium pricing strategy.

Financial contracting between firm and bank

Suppose that the firms have to take up debt to finance their fixed investments. Then, the optimal contract maximizes the expected net profits of each firm subject to *(i)* the incentive-compatibility constraint which ensures truthful reporting of profits, *(ii)* the entrepreneur's limited liability constraints, and *(iii)* the bank's participation constraint. The firm's maximization program is given as:

$$\max_{R_1^H R_1^L \beta^H \beta^L} V = \theta\left[\Pi_1^H - R_1^H + \beta^H (E(\Pi_2) - R_2^H)\right] + (1-\theta)\left[\Pi_1^L - R_1^L + \beta^L (E(\Pi_2) - R_2^L)\right],$$

(3.30)

subject to

$$\Pi_1^H - R_1^H + \beta^H (E(\Pi_2) - R_2^H) \geq \Pi_1^H - R_1^L + \beta^L (E(\Pi_2) - R_2^L),$$

(3.31)

$$\Pi_1^L \geq R_1^L; \quad \Pi_1^L - R_1^L + \Pi_2^L \geq R_2^L,$$

(3.32a)

$$\Pi_1^H \geq R_1^H; \quad \Pi_1^H - R_1^H + \Pi_2^L \geq R_2^H,$$

(3.32b)

$$\theta[R_1^H + \beta^H (R_2^H - I_2)] + (1-\theta)[R_1^L + \beta^L (R_2^L - I_2)] - I_1 \geq \overline{W}.$$

(3.33)

Condition (3.31) is the incentive-compatibility constraint. The limited liability constraints (3.32) state that the firm cannot be forced to pay out more than the profits in the two periods. In the participation constraint of the bank (3.33), $\overline{W}$ reflects the expected gross profit of the bank. If $\overline{W} \geq 0$, this can be interpreted as an indicator of the market power of banks in the financial sector. Strictly positive minimum profit levels $\overline{W}$ for the bank induce an additional source of imperfection in the capital market. This is captured in the present model.

To solve the entrepreneur's maximization problem, we first note that the bank's participation constraint must be binding in the optimum: If not, R_1^H could be lowered without violating (3.33), and the entrepreneur could enjoy higher net first-period profits. By substituting this equation into (3.30), we obtain the entrepreneur's modified objective function as

$$\max V = E(\Pi_1) - I_1 + (\theta \beta^H + (1-\theta) \beta^L)[E(\Pi_2) - I_2] - \overline{W}. \quad (3.30')$$

Moreover, the incentive-constraint is binding as well. By inserting equation (3.31) into the entrepreneur's modified objective function and rearranging terms, we obtain for the refinancing probability after a low demand-shock

$$\beta^L = \frac{\beta^H \theta[E(\Pi_2) - I_2] - (I_1 - R_1^L) - \overline{W}}{\theta[E(\Pi_2) - I_2] + (I - R_2^L)}. \quad (3.34)$$

We see that the refinancing probabilities β^H and β^L are positively related to each other. Maximization of (3.30') implies maximal values for R_1^L and R_2^L. The limited liability constraints (3.32a) are binding, too. With $\beta^H = 1$ we obtain

$$\beta^L = \frac{\theta[E(\Pi_2) - I_2] - (I_1 - \Pi_1^L) - \overline{W}}{\theta[E(\Pi_2) - I_2] + (I - \Pi_2^L)}. \quad (3.35)$$

Using the results $R_1^L = \Pi_1^L$, $R_2^L = \Pi_2^L$, $\beta^H = 1$ and (3.35), we derive from the incentive compatibility constraint that the sum of repayments in good states of nature, i.e. after a high demand shock, must equal

$$R_1^H + R_2^H = \Pi_1^L + E(\Pi_2) - \beta^L[E(\Pi_2) - \Pi_2^L], \quad (3.36)$$

with $\Pi_1^H \geq R_1^H > \Pi_1^L$. Thus, it is not the individual values of R_1^H and R_2^H that affect the constraints and the objective function, but the sum of both values. To summarize, the optimal contract specifies $\{R_1^L = \Pi_1^L, R_2^L = \Pi_2^L, \beta^L$ as in (3.35), $\beta^H = 1$, $R_1^H + R_2^H$ as in (3.36)$\}$. Substituting back into (3.30') yields the firm's expected value under debt financing

3.3 Credit financing and product market competition

$$V = E(\Pi_1) - I_1 + (\theta + (1-\theta)\beta^L)[E(\Pi_2) - I_2] - \overline{W}, \quad (3.37)$$

with β^L as specified in (3.35). If $\theta[E(\Pi_2) - I_2] - (I_1 - \Pi_1^l) - \overline{W} > 0$, then the refinancing probability β^L is strictly positive. This positive refinancing probability in case of low profits, $\beta^l > 0$, differs from the result of Bolton and Scharfstein (1990). In their model, banks have all the bargaining power and, therefore, can completely deny refinancing when profits are low, i.e. $\beta^L = 0$.

Impact of the financial contract on product market competition

How does the financial contract influence the firms' pricing strategies? We insert the results of the product market competition game into the firm's value under debt financing (3.37) and obtain

$$V_i = (p_i^1 - c)\bar{z}\left[\frac{T + p_j^1 - p_i^1}{2T}\right] - I_1$$

$$+ [\theta + (1-\theta)\beta^L](p^r - c)\bar{z}\left[\frac{T + p_j^1 - p_i^1}{2T}\right] - I_2 - \overline{W}, \quad (3.38)$$

with β^l determined in the financing contract as specified in (3.35) before price competition takes place. From the first-order conditions we derive the price reaction functions for firm i and firm j. The symmetric first-period prices under debt-financing are then given by

$$p_{d.f.}^1{}^* = T + c - [\theta + \beta^L(1-\theta)](p^r - c). \quad (3.39)$$

By comparing this to the first-period prices when both firms are self-financed (3.29), we see that equilibrium prices are higher under debt financing. The reason for this is that under debt financing firms have less incentives to invest in market share, because with positive probability they cannot reap the benefits of investing in loyal customers. This result corresponds to the findings of Chevalier and Scharfstein (1996) and Dasgupta and Titman (1998).

Impact of the banks' market power

In case of debt financing, market power in the banking sector reduces the probability of refinancing. From

$$\beta^L = \frac{\theta[E(\Pi_2) - I_2] - (I_1 - \Pi_1^l) - \overline{W}}{\theta[E(\Pi_2) - I_2] + (I - \Pi_2^L)}, \quad (3.35)$$

we derive that an increase in $\overline{W}$ has a direct negative effect on the refinancing probability β^L. In addition to this, there is an indirect effect working through Π_1^L: By inserting the equilibrium price under debt financing into the low first-period profits, we obtain

$$\Pi_1^L(z^L) = z^L(p_i^1 - c)\left[\frac{1}{2} + \frac{p_j^1 - p_i^1}{2T}\right] = z^L[T - (\theta + (1-\theta)\beta^L)(p^r - c)]/2. \quad (3.40)$$

It can easily be seen that $\partial \Pi_1^L / \partial \beta^L < 0$. The indirect effect actually raises the value of (3.35). The overall effect of a bank's market power on β^L is, however, negative. This implies that market power of banks reduces the refinancing probability and hence the efficiency of the financial contract. Moreover, the greater the banks' market power, the smaller is the firms' strategic price cut in the first period in order to invest in market share. In this sense, imperfect competition in the banking sector softens price competition in the product market.

Discussion

Stadler presents a two-period model of product market competition where firms are financed with a debt contract in the design of Bolton and Scharfstein (1990). The financial contracts are composed of two crucial features. The first one is a fixed repayment from the entrepreneur to the lender. The contract thus resembles a standard debt contract. The second one is the probability that the lender grants financing for another period of competition. The bank's threat to stop financing after the first period of competition mititgates the entrepreneur's moral hazard problem. Ex post, repayments and refinancing probabilities have to be contingent on the announcement of the state of nature, i.e. the entrepreneur's report of profits after a high or low demand shock. In contrast to Bolton and Scharfstein (1990) however, the bargaining power in financial contracting is on the side of the firms. This implies that the firm will in general obtain better terms of contract than when the bank has all the bargaining power. Here, it results in a higher refinancing probability for the firm in bad states of nature.

As far as the interaction between financial markets and product market is concerned, the financial contract changes a firm's competitive behavior in the product market. The reduced financing probability induces the firm to operate more short-term orientated: Under debt-financing, a firm invests less in the acquisition of market shares and less in price reduction. Thus, when both firms are financed via a debt contract, product prices are higher than under self-financing. This stands in accordance with the theoretical findings of Dasgupta and Titman (1998) and Chevalier and Scharfstein (1996). The result stands also in line with the empirical findings of Chevalier (1995a,b), Phillips (1995), and Kovenock and

3.3 Credit financing and product market competition

Phillips (1995), although these studies do not test the features of financial contracts, but only the impact of debt-equity ratios on product pricing strategies.

Moreover, Stadler considers market power of banks as an additional variable in financial contracting. This is especially important when the banking sector is not competitive, but is dominated by a few powerful banks, as one might describe the commercial- and house-banking sector in Germany. Market power of banks between monopoly and perfect competition is also analyzed by Ramser and Stadler (1995). Here, Stadler shows that market power of banks reinforces capital market imperfections caused by information problems in financial contracting. Stadler demonstrates that these financial market imperfections, in turn, spill over to the product market and entail an increase in product prices.

Although Bolton and Scharfstein contracts have the attractive feature of mitigating moral hazard problems without costly state verification, they may be criticized for β-values between zero and one. For refinancing probabilities between zero and one, there must exist an enforceable randomizing scheme in order to guarantee feasibility of the financial contract. Of course, no bank tosses a dice in order to decide whether to grant refinancing or not.

Faure-Grimaud (2000) here offers an alternative explanation: He interprets the refinancing mechanism as a debt contract with collateral, i.e. the firm has pledged an asset. Then, β-values between zero and one correspond to a partial liquidation of the pledged asset. The asset's market value is I_2, while its firm-specific value equals $E(\Pi_2)$. If the firm reports low first-period profits Π_1^L and makes the corresponding repayment R_1^L, the entrepreneur will receive $E(\Pi_2)$ with probability β^L, while the portion $1-\beta^L$ of the asset is liquidated. The main result of Faure-Grimaud (2000) is the same as above: This kind of two-period debt contracts induce less aggressive pricing strategies of firms in the product market.

One last remark concerns the nature of the demand shocks. What happens if the demand parameters are not the same for both firms, but are firm-specific? If one firm realizes a high z, while the other firm obtains a low z-value, the low-z firm will with a certain probability receive no refinancing and, thus, will have to drop out of the market. This, in turn, raises industry concentration. We will examine these impacts of financial contracting on the structure of an industry in the following subsection.

3.3.2 Credit financing, innovation, and product market competition

> *Innovation is an activity where "dry holes" and "blind alleys" are the rule, not the exception.*
> Oz Shy (1995)

3.3.2.1 Introduction

We present a two-period model of innovation and strategic competition where innovation activities depend on the competitive market position of firms. Firms in need of external funding are financed by credit contracts in the design of Bolton and Scharfstein.

Our model extends the existing literature in the following ways:

(i) In order to capture product market dynamics, we develop a two-period version of price competition in the circular city-model (Salop 1979), instead of using the rather static consumer switching cost approach.

(ii) In addition to price competition, the firms engage in R&D activities, because they wish to realize cost-reducing process innovation. These R&D activities are extremely important in highly competitive industries, where firms fiercely compete for market shares, e.g. in the computer hardware or the printing press industry, but also in the automotive, industrial manufacturing, and electronics industries.

(iii) In our model, firms must make decisions under uncertainty. In contrast to Chevalier and Scharfstein (1996) or Stadler (1997), however, this uncertainty doesn't stem from stochastic product demand, but is created by the unknown result of innovation activities.

(iv) Concerning the financial contracts, we make the standard assumption that banks are less informed about the firms' profits than entrepreneurs. Since banks do not have a costly state verification technology at their disposal, they will threaten not to refinance the firm if the entrepreneur defaults on her repayment obligation. In this way, the financial contracts resemble the Bolton and Scharfstein (1990) framework. The design of our contracts, however, is conditioned on the facts that *(a)* we allow profits to vary across periods, and

3.3 Credit financing and product market competition 77

that *(b)* we allow firms to acquire non-symmetric market positions.[4] This implies that our financial contracts will become much more complex.

(v) We then will investigate the impact of financial contracting on product market competition if either one or both of the firms need external funding. We analyze how the individual financing decision interacts with the firms' R&D expenditures, their pricing policies, and the competitive structure of the industry. Moreover, we show that credit-financing makes a firm vulnerable for predatory activities from part of a self-financed rival.

The section is organized as follows: We first analyze a two-period game of innovation and price competition when both firms are self-financed (subsection 3.3.2.2). We then investigate the case that one of the firms depends on outside financing. We derive the optimal credit contract and analyze its impact on the innovative activities and the competitive behavior of both firms (3.3.2.3). Next, we ask how innovation dynamics, product market competition and the resulting market structure will change if both firms need external debt financing (3.3.2.4). Lastly, we address the subject of predation (3.3.2.5). The analysis concludes with a discussion of the results.

3.3.2.2 A dynamic game of innovation and price competition

In the present section we analyze a simple model of cost-reducing innovation and competition when both firms are able to finance R&D activities and fixed production costs from internal resources.

Assumptions

A1: Players: There are two firms *i, j* that compete in the product market. Competition is in prices with horizontally differentiated products. The firms have sufficient funds to self-finance their investments.

A2: The model has two time periods, period 1 and period 2. In each period the firms must pay a fixed amount I_t before operating in the market. The competition game consists of two stages: First, the firms decide on how much R&D expenditures they want to spend in the present period. Nature decides whether the innovation activities will be successful or not. In a second step, both firms choose their pricing strategies according to the result of the innovation activities.

A3: Information structure: The results of the innovation process becomes known to all players. Whether there has been a successful innovation, i.e. cost-

---

[4] Maurer (1998) also investigates Bolton and Scharfstein contracts when firms have non-symmetric market positions. In his model, however, profits are exogenously given, and competition in the product market is not explicitly formalized.

A4: The firms are risk-neutral.

A5: The firms produce with constant marginal costs c. Innovations lead to a stagewise reduction of marginal costs: If a firm successfully innovates in one of the (both) periods, marginal costs will decrease from a high to a middle (low) level: $c_h \to c_m \to c_l$. We assume that the cost differentials $\Delta c = c_h - c_m = c_m - c_l$ are symmetric.

Innovations are induced by firm-specific R&D activities. These R&D activities determine the probability of success θ for each innovation project. The innovation project is associated with additional costs $g(\theta)$ to the firm. We assume these costs to be increasing and convex in the success probability, $g'(\theta) > 0$, $g''(\theta) > 0$. Thus, the higher the research expenditures $g(\theta)$, the higher the success probability of the innovation project θ. For simplicity we specify the research costs to be quadratic, i.e. $g(\theta) = \rho\theta^2$ with $\rho \in R^+$.[5]

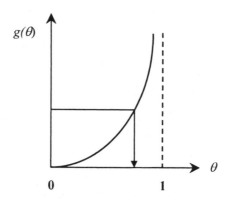

Figure 3.4: Research costs and innovation success probability

Before production in each period starts, firms incur fixed production costs $I_1 = I_2 = I_t$. The expected value of investment in each period is positive, $I_t \le \overline{\Pi}^i - g(\theta)$. We further assume that the market offers just enough space for two differentiated products, which means that any potential entrant would make negative profits.

---

[5] We assume the parameters to be such that an interior solution exists; otherwise we assume that $g(\theta)$ is approaching infinity at $\theta = 1$, i.e. that research costs are too high to induce innovation with certainty.

3.3 Credit financing and product market competition

A6: Consumers have different tastes for the product variants supplied. In order to explicitly capture product market dynamics, we change the circular city-model of product differentiation into a dynamic model of repeated product market interaction.[6] Our dynamic model grasps the fact that in dynamic technology-intensive markets, the incentives to innovate arise from period-specific, transitory market power: Firms make investments in order to gain market shares. Very important is that these market shares are not fixed as in Klemperer's (1987) consumer switching cost approach, but that they can be gained and also lost. This implies that firms have to defend them and, therefore, actively engage in R&D activities. Firms here have to compete in both time periods for their customers.[7]

Consumers are uniformly distributed along a circle with a perimeter of one. The firms are located exactly opposite to each other on this circle. Consumers have a reservation value $\bar{s}$ for each unit they purchase. In addition to the price, they bear a cost of T per unit of distance to the firm of their choice. This cost measures the utility loss a consumer incurs when her preferred variant of product is not supplied. Preferences are such that each consumer buys one good per period of time. A consumer at point x buys from firm i as long as the sum of price and utility loss is smaller than at firm j:

$$p_i + T \cdot x \leq p_j + T \cdot (0,5 - x)$$

The indifferent consumer determines the firm-specific demand. Since there is an indifferent consumer on either side of the firm's location, the total market share of company i equals:

$$\Leftrightarrow \quad 2x_i = \frac{T/2 + p_j - p_i}{T} \tag{3.41}$$

We see that market share increases in the rival's price and decreases in the own price.

---

[6] This stands in contrast to the consumer ‚switching cost' approach of Klemperer (1987), in which firms compete only in the first period actively for customers, whereas in the second period market shares stay the same because high transaction costs prevent consumers from switching to another product variant.

[7] A nice example for repeated competition in the sense of our model is the flowmeter industry (see Sutton 1998, chapter 6). In the flowmeter industry, horizontal product differentiation is created by the physical principle employed in measurement, e.g. there are electromagnetic, ultrasonic, Coriolis force flowmeters. In the 1980s, technological competition between electromagnetic and ultrasonic flowmeters was very intense. Innovations lead to substantial cost-reduction for each product type. Firms took alternate turns in successful innovation, price cuts and the acquisition (the loss) of market shares.

80   3 Credit financing and strategic competition

A7: There is no financial contracting, because firms are able to self-finance all their investment expenditures.

The basic game

The objective function of firm i (for firm j analogously) in the basic game consists of the following value function:

$$V_i^1 = \overline{\Pi} - I_1 - g(\theta_i) = (p_i - c)(\frac{T/2 + p_j - p_i}{T}) - I_1 - g(\theta_i), \qquad (3.42)$$

Besides the gross profits which are given by the price-cost margin times the quantity sold, it comprises the fixed costs for overall production I_1 as well as R&D expenditures $g(\theta)$.

The optimal prices are set according to the first order condition $\partial \overline{\Pi}_i / \partial p_i = 0$. We solve for the (own) price and obtain the reaction function. The Nash-equilibrium lies at the intersection of these reaction functions. Depending on the results of the innovative activities we derive the following prices and corresponding gross profits for each firm in the first period:

Situation	Price	Market Share	Gross profit
(11) both firms are successful	$p_i^{11} = \frac{T}{2} + c_m$	$\sigma_i = 50\%$	$\Pi_i^{11} = \frac{T}{4} = \Pi_i^S$
(10) only firm i is successful	$p_i^{10} = \frac{1}{2}T + \frac{1}{3}c_h + \frac{2}{3}c_n$	σ_i increases	$\Pi_i^{10} = \frac{1}{4}T + \frac{1}{3}\Delta c$ $+\frac{1}{T}(\Delta c/3)^2 = \Pi_i^A$
(00) both firms fail	$p_i^{00} = \frac{T}{2} + c_h$	$\sigma_i = 50\%$	$\Pi_i^{00} = \frac{T}{4} = \Pi_i^S$
(01) only firm i fails	$p_i^{01} = \frac{1}{2}T + \frac{2}{3}c_h + \frac{1}{3}c_n$	σ_i decreases	$\Pi_i^{01} = \frac{1}{4}T - \frac{1}{3}\Delta c$ $+\frac{1}{T}(\Delta c/3)^2 = \Pi_i^D$

Table 3.1: Prices, market shares, and profits after the first period of innovation

We concentrate on firm i since prices and profits are mutually symmetric. Profits are highest when a firm successfully innovates and realizes a cost-reduction, while the rival does not. These profits are influenced positively by the difference between marginal costs, Δc. The following relation holds for the first-period profits:

$$\Pi_i^{10} > \Pi_i^{11} = \Pi_i^{00} > \Pi_i^{01}.$$

We therefore can write: $\Pi^A > \Pi^S = \Pi^S > \Pi^D$, where Π^A stands for the gross profits for the firm with the cost-advantage, Π^S stands for the gross profits in the symmetric cases, and Π^D stands for the gross profits of the firm with the cost-disadvantage. We see that profits are equal in the symmetric cases, $\Pi^{11} = \Pi^{00}$. Thus, the model demonstrates that profits are not based on the absolute but on the relative performance, i.e. the market position, of a firm.

R&D Activities:

In order to determine the optimal level of R&D activities, we calculate the expected value of the first-period profits.

$$V^1 = \theta_i\theta_j\Pi_i^S + \theta_i(1-\theta_j)\Pi_i^A + (1-\theta_i)(1-\theta_j)\Pi_i^S + (1-\theta_i)\theta_j\Pi_i^D - g(\theta_i) - I_1 \quad (3.43)$$

The firm chooses the success probability θ_i which maximizes the expected per-period profit less the R&D expenses $g(\theta_i)$ and the fixed operating costs I_1. The first-order condition to this problem is give as

$$g'(\theta_i) = \theta_j \cdot (2 \cdot \Pi_i^S - \Pi_i^A - \Pi_i^D) + \Pi_i^A - \Pi_i^S. \quad (3.44)$$

In the optimum, marginal R&D expenditures must equal expected marginal profits from innovation. The reaction functions of the R&D activities exhibit strategic substitutability[8]:

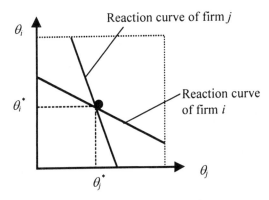

Figure 3.5: R&D reaction functions in the basic game

---

[8] With research costs specified as $g(\theta) = \rho\theta^2$, we derive linear reaction functions. Research cost functions of third order work as well, but complicate the analysis tremendously.

In the symmetric equilibrium $\theta_i = \theta_j = \theta^*$ and therefore

$$\theta_i^* = \frac{\Pi_i^A - \Pi_i^S}{2\rho + [(\Pi_i^A - \Pi_i^S) - (\Pi_i^S - \Pi_i^D)]} = \theta_j^* \qquad (3.45)$$

with $2\rho = g'(\theta)/\theta$.

The optimal level of the R&D activities mainly depends on the difference between the profits of a firm with a cost-advantage and the profits in the symmetric case $\Pi^A - \Pi^S$. The term 2ρ in the denominator of (3.45) is influenced by the slope of the R&D-cost function and has a negative impact on the optimal R&D level: The more "expensive" the research costs, the lower the optimal success probability.

The two-period case

We now extend our model to two periods of innovation and product market competition. The timing is as follows.

In $t=1$ the two symmetric firms decide upon their R&D activities. Nature determines the success of the innovation projects. The firms choose their optimal prices contingent on the marginal costs, the products are sold, and profits are realized. The first period is equivalent to the one-period case.

In $t=2$ the competitive position of a firm is determined by the innovation results of the first period. There are three possible starting positions for each firm:

- The firms have a symmetric market positions, i.e. either both firms have innovated or none of them has.
- Firm i starts with a cost advantage, because it has been the only innovator in the previous period.
- Firm i starts with a cost-disadvantage because the rival has been the only innovator in the first period.

Depending on their relative market position, the firms again simultaneously choose their R&D levels. By μ we denote the R&D activities of the second period, μ_i and μ_j. Nature determines whether the second-period R&D efforts will be successful or not. Then the firms learn their marginal costs of production and set the optimal prices accordingly. Consumers buy the products supplied and profits are realized.

Again, we solve for the subgame-perfect Nash-equilibrium of this two-period multi-stage game by backward induction.

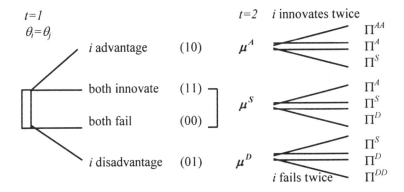

Figure 3.6: Innovation success probabilities and payoffs in the two-period case

Second period

By analyzing the profit opportunities of the second period, we see that the dynamic circular city-model reports equal profits for equal market positions. Altogether we have five different profit situations: Π^S again indicates that firms have symmetric market positions after the second round of R&D activities and product market competition. Π^A represents the second-period profit situation for a firm with a single cost-advantage and Π^{AA} represents a double cost-advantage. Π^D on the other hand denotes that the firm has a cost-disadvantage, and Π^{DD} means that the firm still produced with high costs in the second period while its rival has successfully innovated twice. Note that leapfrogging from a high directly to a low cost-level is not possible

We specify μ^S as the level of R&D activities for the firms in the symmetric cases; and μ^A, μ^D represent the R&D levels for a firm that realized a cost-advantage or a disadvantage in the previous period, respectively.

We begin by analyzing the symmetric cases:

If none of the firms innovates in the first period, decision-making in the second period will be a mere repeating of the situation in the „isolated" first period. Pricing strategy and profit possibilities are as shown as in Table 3.1. R&D activities are identical: $\mu^S = \theta^*$. The expected gross profit in the symmetric case is $\overline{\Pi}_S^2$, from which $g(\mu^S)$ and I_2 have to be deducted. This scenario occurs with probability $(1-\theta_i)(1-\theta_j)$.

If both firms successfully reduced marginal costs in the first period, the situation is quite similar. Since their competitive position is symmetric, incentives to further innovate are again $\mu_i = \mu_j = \mu^S = \theta^*$. Expected net profits for both firms likewise amount to $\overline{\Pi}_S^2 - g(\mu^S) - I_2$ with the only difference that the product prices are lower now. This scenario occurs with probability $\theta_i \theta_j$.

More interesting are the cases with asymmetric starting positions. If firm i failed to innovate in the first period while its rival j was successful, then in the second period firm i will only regain its competitive position if it innovates and the rival does not. After setting the optimal prices, the profit situation for the firm with the cost-disadvantage is as follows:

$t=2$: Firm i starts with a cost-disadvantage	j succeeds	j fails to innovate
i succeeds	$\Pi_i^D = \frac{1}{4}T - \frac{1}{3}\Delta c + \frac{1}{T}(\Delta c/3)^2$	$\Pi^S = \frac{T}{4}$
i fails	$\Pi_i^{DD} = \frac{1}{4}T - \frac{2}{3}\Delta c + \frac{1}{T}(\frac{2}{3}\Delta c)^2$	$\Pi_i^D = \frac{1}{4}T - \frac{1}{3}\Delta c + \frac{1}{T}(\Delta c/3)^2$

Table 3.2: Second-period gross profits if firm i starts with a cost-disadvantage

Profits are highest when i regains the lost market shares and realizes Π^S. Profits are lowest when i fails to innovate for a second time and must face a double cost-disadvantage, which results in Π^{DD}. If both firms innovate or both fail, the cost-disadvantage for i remains and the firm realizes profits Π_i^D.

In contrast to this, the single innovator of period 1 has always better profit possibilities: In the second period, the firm can keep its cost advantage (if either both firms innovate or none of them does), or can even enhance it, if the firm succeeds while the rival fails to innovate a second time. We consider the second-period profit situation for the firm with a cost-advantage in Table 3.3 below:

$t=2$: Firm i starts with a cost-advantage	j succeeds	j fails to innovate
i succeeds	$\Pi_i^A = \frac{1}{4}T + \frac{1}{3}\Delta c + \frac{1}{T}(\Delta c/3)^2$	$\Pi_i^{AA} = \frac{1}{4}T + \frac{2}{3}\Delta c + \frac{1}{T}(\frac{2}{3}\Delta c)^2$
i fails	$\Pi^S = \frac{T}{4}$	$\Pi_i^A = \frac{1}{4}T + \frac{1}{3}\Delta c + \frac{1}{T}(\Delta c/3)^2$

Table 3.3: Second period gross profits if firm i starts with a cost-advantage

We see that the profits of the cost-leader are positively influenced by the difference of the marginal costs between her and her rival. Here, profits are highest in the case of repeated single innovation by firm i, which results in a double cost-advantage and associated profits Π^{AA}.

3.3 Credit financing and product market competition

Taking the expected profits into account we now can derive the optimal second-period R&D activities for the firm with a cost-disadvantage and a cost-advantage, respectively:

$$\mu^D = \frac{\Pi^S - \Pi^D + (2\cdot\Pi^A - \Pi^S - \Pi^{AA})(\Pi_j^{AA} - \Pi_j^A)}{2\rho - (2\cdot\Pi^D - \Pi^S - \Pi^{DD})(2\cdot\Pi_j^A - \Pi_j^S - \Pi_j^{AA})} \quad (3.46)$$

$$\mu^A = \frac{\Pi^{AA} - \Pi^A + (2\cdot\Pi^A - \Pi^S - \Pi^{AA})(\Pi_j^S - \Pi_j^D)}{2\rho - (2\cdot\Pi_j^D - \Pi_j^S - \Pi_j^{DD})(2\cdot\Pi^A - \Pi^S - \Pi^{AA})} \quad (3.47)$$

with $2\rho = g'(\mu)/\mu$.

We see that the optimal level of R&D activities is influenced by the profit values of both firms.[9] The denominators in both formulas are the same. The main factor that determines the level of optimal research activities are the additional profits resulting from a cost-difference in period two: $(\Pi^S - \Pi^D)$ for the firm with the cost-disadvantage and $(\Pi^{AA} - \Pi^A)$ for the firm with the cost-advantage.

Proposition 3.5 (Second-period R&D investments)

(i) For the optimal level of R&D activities in the second period, the following relation holds:

$$\mu^A > \mu^D.$$

The cost-leader will invest more in R&D activities than the cost-follower.

(ii) Moreover, if potential gains from innovation are very high, firms in symmetric market positions will compete more fiercely in second-period R&D game than firms in asymmetric market positions:

$$\mu^S > \mu^A > \mu^D.$$

The proposition states that a previously successful firm will invest more in second-period R&D activities than an unsuccessful rival. The reason is that a firm with a double cost-advantage can lower its price substantially and, thus, can convince an even larger part of consumers to buy its product variant.[10] The fact that the cost-leader will invest more in R&D activities than the cost-follower

---

[9] For notational simplicity we suppress the index i and mark only the rival's influence by index j.

[10] See Appendix 3.5 for a detailed analysis of the influence of Δc, T, and ρ on the optimal second-period R&D levels μ.

reflects the hypothesis of "success breeds success". On the other hand, a firm that has fallen behind in the first-period innovation game can at most regain its lost market share, but cannot acquire new customers. This is due to the fact that leapfrogging is not possible. Nevertheless, this firm has to engage in R&D activities too in order not to fall behind a second time.

Moreover, by comparing the R&D reaction functions for the symmetric and the asymmetric cases, we derive that firms in symmetric market positions will spend more on R&D activities, i.e. $\mu^S > \mu^A$, if the parameter of the research cost function ρ is lower than one third of the potential decrease in production costs, $\rho < (\Delta c)/3$, (see Appendix 3.5).

With the optimal values for μ^A, μ^D and μ^S we derive the expected second-period gross profits for the firm with the cost-advantage $\overline{\Pi}_A^2$, the firm with the cost-disadvantage $\overline{\Pi}_D^2$, and for firms in symmetric market positions $\overline{\Pi}_S^2$.

First period

When applying backward induction, we take into account that firms will anticipate the expected second-period profits while choosing their optimal first-period R&D activities. The optimal R&D activity of the first period is determined by including the expected second-period profits into the maximization program. We obtain for the equilibrium R&D activities under the long-term horizon (^{LT}):

$$\theta^{LT} = \frac{\Pi^A + [\overline{\Pi}_A^2 - g^A] - \Pi^S - [\overline{\Pi}_S^2 - g^S]}{2\rho + \Pi^A + \overline{\Pi}_A^2 - g^A - \Pi^S - \overline{\Pi}_S^2 + g^S - \Pi^S - \overline{\Pi}_S^2 + g^S + \Pi^D + \overline{\Pi}_D^2 - g^D},$$

with $2\rho = g'(\theta)/\theta$; $g^S = g(\mu^S)$, $g^A = g(\mu^A)$, $g^D = g(\mu^D)$. (3.48)

Graphically, this causes the R&D reaction functions to shift outward, as can be seen in Figure 3.7:

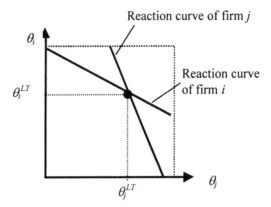

Figure 3.7: R&D reaction functions in the two-period game

Proposition 3.6 (First-period R&D investments in the dynamic game)

The optimal R&D activities in the first period will increase, if the second period of competition is taken into account:

$$\theta^{LT} > \theta^*.$$

This result is consistent with the intuition that if there are higher gains from innovation due to the repeated product market interaction, firms will spend more on first-period R&D expenditures. If firms anticipate that future profits in the second period are strictly positive, and depend on the market position realized after period one, the firms will strategically increase their R&D expenditures in the first period.

This result corresponds to the findings of Chevalier and Scharfstein (1996) which state that strategic investments will increase if firms interact more than once in the product market. While in their model firms directly invest in strategic price cuts, the firms here strategically invest in R&D activities, which reduce expected marginal production costs, which in turn indirectly induces strategic price cuts.

3.3.2.3 Optimal debt contract for one externally financed firm

In this subsection we now will analyze innovation activities and product market competition if one of the firms needs external debt financing. We assume that capital markets are imperfect due to asymmetric information. This gives rise to a moral hazard problem between the entrepreneur and the bank. Since costly state verification is not possible, the bank offers only long-term contracts in the design of Bolton and Scharfstein.

We first present the modified assumptions, then derive the optimal financial contract between the firm and its bank. Finally, we investigate the changes in the innovation activities and the competitive structure of the industry which are induced by the financial contract.

Assumptions

A1: Players: Two firms i, j compete with each other in the heterogenous product market. Firm i does not have sufficient internal funds to finance its fixed production costs, and seeks financing from a bank. The banking sector is competitive. The representative bank has all the bargaining power and offers a take-it-or-leave-it long-term debt contract to firm i. Firm i is subject to limited liability.

A2: The time horizon is two periods. The financial contract is offered and signed before the first period of innovation and competition takes place.

A3: Information structure: If firm i applies for a bank loan, an information problem arises between the firm and the bank because of the stochastic profit situation. Although the bank is perfectly informed about the competitive structure of the game, i.e. the profit possibilities in the first and second period, and the actual R&D costs, the bank is unable to observe which event (success or failure) nature assigns to the innovative efforts of firm i. Furthermore, the bank cannot observe product prices either: Only a firm operating in the market knows the exact price constellations and is informed about the innovation success or failure of its competitor. As a consequence, the bank does not know which profit level firm i has finally realized. Thus, firm i has always an incentive to lie and to announce low profits in order to reduce repayments of the loan.

A4: All players are risk-neutral.

A5: Production technology: We point out again that the expected value of firm i's investment in each period is positive, $I_t \leq \overline{\Pi}^I - g(\theta)$. Thus, if firm i could finance itself through internal funding, it would invest in each period. To make things interesting, however, we assume that fixed production costs I_t are higher than net profits of a firm with a cost-disadvantage, $I > \Pi^D - g(\theta)$. With positive probability firm i thus will not be able to repay its loan.

A6: Consumer demand is specified as in the dynamic circular city-model above.

3.3 Credit financing and product market competition

A7: Financial contracting: As in Bolton and Scharfstein (1990) we assume that the bank has all the bargaining power in financial contracting. In reality, bargaining power will be distributed between both contracting parties. Our results, however, i.e. the impact of the financial contract on innovation and product market competition, become more distinct in this way.

The amount of credit in each period is I_t, $t=1,2$. We suppose that the firm cannot offset borrowing needs against first-period profits; or, equivalently, that profits net of loan repayments and research expenditures do not suffice to cover fixed production cost in the second period, $\Pi^A - g(\theta) - R^A < I_2$.

The contract is based on the entrepreneur's reported profits. If the entrepreneur announces first-period profits to be low, the bank will select its refinancing probability for the second period accordingly. For incentive reasons, the bank has even the right to completely deny follow-up financing for the second period of competition.

The optimal financial contract

The contract is a direct revelation mechanism. It is based on reported profits Π^D, Π^S, Π^A, and specifies the first-period repayments R^D, R^S, R^A, the respective refinancing probabilities β^D, β^S, β^A, and the repayments in the second period, R^2. Notice that the second-period repayments are all identical because of the moral hazard problem: The entrepreneur will always announce low profits. The bank has no threat potential to elicit truthful reporting in the second period, because the firm doesn't need further refinancing after $t=2$.

The optimal financing contract maximizes the expected profits W_i of the bank subject to *(i)* the limited liability constraints *(ii)* the condition for the optimal R&D activities of the firm, and *(iii)* the incentive compatibility constraints which ensure that the entrepreneur reports her profits truthfully.

The maximization problem of the bank can be formulated as follows:

$$\max W = \theta_i \theta_j R^S + \theta_i (1-\theta_j) R^A + (1-\theta_i) \theta_j R^D + (1-\theta_i)(1-\theta_j) R^S - I_1$$
$$+ [\beta^S \theta_i \theta_j + \beta^A \theta_i (1-\theta_j)](R^2 - I_2) \qquad (3.49)$$
$$+ [\beta^D (1-\theta_i) \theta_j + \beta^S (1-\theta_i)(1-\theta_j)](R^2 - I_2).$$

The objective function of the bank (3.49) consists of the expected repayments less the amount borrowed in the first-period, plus the total probability of future finance multiplied with the net repayments to the bank of the second period.

The maximization problem is subject to the following first-period limited liability constraints:

$$\Pi^D - g(\theta_i) \geq R^D, \quad \Pi^{DD,D,S} - g(\mu^D) \geq R^2,$$

$$\Pi^S - g(\theta_i) \geq R^S, \quad \Pi^{D,S,A} - g(\mu^S) \geq R^2, \tag{3.50}$$

$$\Pi^A - g(\theta_i) \geq R^A, \quad \Pi^{S,A,AA} - g(\mu^A) \geq R^2.$$

The limited liability constraints (3.50) imply that repayments cannot exceed the actual profits net of innovation costs.

Moreover, the bank's maximization problem is also subject to the condition for the optimal level of R&D activities:

$$\begin{aligned}
g'(\theta_i) = & \, \theta_j \left[\Pi^S - R^S + \beta^S (\overline{\Pi}_S^2 - g(\mu^S) - R^2) \right] \\
& - \theta_j \left[\Pi^D - R^D + \beta^D (\overline{\Pi}_D^2 - g(\mu^D) - R^2) \right] \\
& + (1 - \theta_j) \left[\Pi^A - R^A + \beta^A (\overline{\Pi}_A^2 - g(\mu^A) - R^2) \right] \\
& - (1 - \theta_j) \left[\Pi^S - R^S + \beta^S (\overline{\Pi}_S^2 - g(\mu^S) - R^2) \right].
\end{aligned} \tag{3.51}$$

Equation (3.51) represents the modified first order condition (3.48) for firm i's optimal R&D activities. The level of R&D activities will change because expected net profits in both periods shift due to the required loan repayments.

Finally, the maximization problem is subject to the firm's incentive compatibility constraints:

$$\Pi^A - R^A + \beta^A (\overline{\Pi}_A^2 - g(\mu^V) - R^2) \geq \Pi^A - R^S + \beta^S (\overline{\Pi}_A^2 - g(\mu^A) - R^2) \tag{3.52}$$

$$\Pi^A - R^A + \beta^A (\overline{\Pi}_A^2 - g(\mu^A) - R^2) \geq \Pi^A - R^D + \beta^D (\overline{\Pi}_A^2 - g(\mu^A) - R^2) \tag{3.53}$$

$$\Pi^S - R^S + \beta^S (\overline{\Pi}_S^2 - g(\mu^S) - R^2) \geq \Pi^S - R^D + \beta^D (\overline{\Pi}_S^2 - g(\mu^S) - R^2) \tag{3.54}$$

The incentive compatibility constraints (3.52), (3.53) and (3.54) ensure that it is always advantageous for the firm to truthfully reveal its actual profits and not to falsely report lower profits instead. This is obtained by granting a different refinancing probability for each reported profit level. Equation (3.53), however, is redundant for the maximization problem, because from comparison with (3.52) and (3.54) we see that the incentive compatibility constraint (3.54) is more restrictive than (3.53).

Solution

To solve the optimization problem we repeat the assumption that the amount borrowed is higher than lowest net first-period profit: $I_1 > \Pi^D - g(\theta)$. The bank therefore incurs losses when firm i fails to innovate while firm j is successful, because profits are too low to cover the required repayment R^D. In this case there

3.3 Credit financing and product market competition

will be no follow-up financing for firm i: $\beta^D = 0$, since firm i cannot credibly commit itself to cover the present losses by higher repayments in the second period. First-period repayments will thus be $R^D = \Pi^D - g(\theta^*) < I_1$.

Because of the moral hazard problem repayments in the second period equal the amount of the smallest profit of the cases with follow-up financing. This is Π^D less the R&D expenditures in the symmetric case, i.e. $R^2 = \Pi^D - g(\mu^S)$. This, however, implies that the bank is always facing a negative return on investment in the second period and has actually to be forced by the terms of contract to extend the second-period loan.

By inserting these results into the incentive compatibility constraints and by taking the limited liability conditions into account we obtain the values for the other repayments and the refinance probabilities: We have to distinguish between two different cases since we have not specified the exact research cost function nor the parameters for the demand side.

Case 1 (Low second-period innovation probability, $\mu^S<0,5$): In this case, the refinance probabilities take the values of $\beta^D=0$ and $\beta^S=\beta^A=1$, and repayments R^S and R^A are identical. Technically speaking, this scenario is valid when expected second-period profits $\overline{\Pi}_S^2$ are smaller than the first-period profits Π^S for equal market position. This condition is derived during the maximization process.

Case 1: Low innovation probability ($\mu^S<0,5$)	$\Pi^D < \overline{\Pi}_S^2 < \Pi^S < \Pi^A$	
Refinance probability	$\beta^D=0$	$\beta^S = \beta^A = 1$
Repayments in $t=1$	$R^D = \Pi^D - g(\theta)$	$R^S = R^A = \overline{\Pi}_S^2 - g(\theta)$
Repayments in $t=2$	--	$R^2 = \Pi^D - g(\mu^S)$

Table 3.4: Optimal contract in case of a low innovation probability ($\mu^S<0,5$).

Case 2 (High innovation success probability, $\mu^S \geq 0,5$): In this slightly more complicated case, the refinance probabilities take the values $\beta^D = 0$ if only firm i fails to innovate in the first period, β^S will lie between zero and one if firms have symmetric market positions, and $\beta^A =1$ if firm i has been the only innovator in the first period. Technically speaking, this scenario is valid when expected second-period profits $\overline{\Pi}_S^2$ are higher than the symmetric first-period profits Π^S.

Case 2: $(\mu^S \geq 0,5)$:	\multicolumn{3}{c}{$\Pi^D < \Pi^S < \overline{\Pi}_S^{t=2} < \Pi^A$}		
Refinance probability	$\beta^D = 0$	$\beta^S = \dfrac{\Pi^S - \Pi^D}{\overline{\Pi}_S^2 - \Pi^D}$	$\beta^A = 1$
Repayments in $t=1$	$R^D = \Pi^D - g(\theta_i^*)$	$R^S = \Pi^S - g(\theta_i^*)$	$R^A \leq R^S + (\beta^A - \beta^S) *$ $(\overline{\Pi}_A^2 - g(\mu^A) - R^2)$
Repayments in $t=2$	--	$R^2 = \Pi^D - g(\mu^S)$	$R^2 = \Pi^D - g(\mu^S)$

Table 3.5: Optimal contract in case of a high innovation probability ($\mu^S \geq 0,5$)

From Table 3.5 we derive that if the symmetric refinancing probability β^S is less than one, the two repayments R^S and R^A will be different. Note, however, that the repayment R^A if firm i is the single innovator exceeds R^S only by a small value due to incentive reasons.

As a next step we insert these results into the condition for the level of optimal R&D activities (3.51). Here, we have to take into account that the R&D activities of the rival firm θ_j, are also affected by the financial contract: If the bank refuses to refinance firm i in the second period, firm i will be obliged to exit the market and firm j will realize monopoly profits in $t=2$. Contingent on the level of these monopoly profits,[11] firm j drastically increases its first period R&D activities. At the same time, firm i reduces its innovation activities because the information problems in financial contracting which are responsible for the reduced refinancing probability, lead to lower expected second-period profits. This induces firm i to cut its innovation activities. As a result we thus derive that the financial contract causes a change of optimal R&D activities for both firms:

Proposition 3.9 (Optimal R&D activities with one externally financed firm)

If firm i is externally financed via a long-term debt contract, its optimal first-period R&D activities will decrease. The self-financed rival j, on the other hand, will increase its level of first-period R&D activities:

$$\theta_i^{d.f.} < \theta^{LT} < \theta_j^{s.f.}.$$

---

[11] The monopoly profits depend on the consumers' reservation values $\overline{s}$. If we assume that the minimum value $\overline{s}$ is just high enough that firm j is able to cover the whole market by asking the original price $p = T + c_h$, then the expected monopoly profits equal $\Pi^{Mono} = T + \Delta c + \mu^{Mono} \cdot \Delta c - g(\mu^{Mono}) - I_2$.

Graphically we see (in comparison to Figure 3.6) that both R&D reaction functions shift as a consequence of the borrowing needs of firm i. The slope and the intercept of the rival's reaction function j is influenced by the size of the monopoly profits. The position of firm i's R&D reaction function depends on the terms of contract: If the financial contract specifies a refinancing probability of $\beta^S = 1$ (Case 1), the shift in firm i's reaction function won't be very drastic:

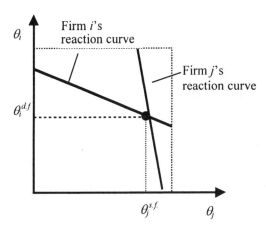

Figure 3.8: R&D reaction functions if firm i is externally financed via a long-term loan contract and the refinancing probability is $\beta^S = 1$ (Case 1)

If, by contrast, the refinancing probability is $\beta^S < 1$ (Case 2), firm i's R&D reaction function will shift toward the origin and will flatten:

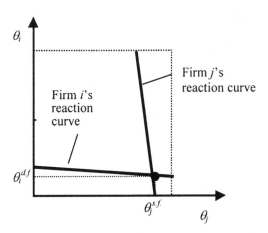

Figure 3.9: R&D reaction functions if firm i is externally financed via a long-term loan contract and the refinancing probability is $\beta^S < 1$ (Case 2).

As a last step, we substitute the results for $\{\theta_i^{d,f}, \theta_j^{s,f}, R^D, R^S, R^A, \beta^D, \beta^S, \beta^A,$ and $R^2\}$ into the bank's maximization problem (3.49), and derive the bank's expected profits W_i. As long as expected profits are positive, the financial contract is actually offered. Since the profits are contingent on the amount of loan extended, $W(I_i)$, the bank will impose an upper bound on I_i.

Impact of the financial contract on innovation and strategic competition

The financial contract of firm i has the following implication for *(i)* the innovation activities, *(ii)* the pricing strategies of both firms, and *(iii)* for the market structure of the industry:

(i) The innovative activities of the financially restricted firm decrease remarkably in the first-period of competition. The reason is that large parts of the firm's profits are transferred to the bank via the financial contract. However, the expected profits of this firm remain positive: Though all the bargaining power is on the side of the bank, the bank is not able to reduce firm i's profits to zero because of the limited liability constraint and the moral hazard problem.

On the other side, innovation efforts of the rival firm will dramatically increase: The rival anticipates the chance to realize monopoly profits if firm i is denied follow-up financing. This induces firm j to innovate with a high probability in the first period of competition. In the second period, firm j attempts to innovate again in order to increase its monopolistic price-cost margin.

(ii) Moreover, expected product prices of the leveraged firm will be higher while those of the self-financed rival will be lower than in the all-equity case. Welfare effects of the first period could therefore be positive.

(iii) But if firm i fails to innovate and doesn't obtain refinancing, the unleveraged rival j becomes a monopolist in the second period. The monopolist will raises its price up to the consumers' reservation value (net of transportation costs) and overall welfare will decline.

3.3.2.4 Optimal contracts when both firms need external debt financing

In this subsection we suppose that not only firm i, but also firm j seeks external debt financing.

Assumptions

A1: Players: In addition to firm i, its bank, and firm j, a second bank participates in the game: The "house"-bank of firm j likewise offers a take-it-or-leave-it long-term debt contract to firm j. Moreover, we assume that no other firm will enter the product market if firm i or j becomes a monopolist in the second period.

A2: The time horizon is two periods. The financial contracts are offered simultaneously at the beginning of the game.

A3: The information structure is the same as in the preceding section: Banks cannot observe product prices nor profits. Because of the moral hazard problem, a bank denies any follow-up financing if its firm falls behind in the innovation game. Since this may happen either to firm i or to firm j, there are two scenarios where a monopolistic market structure arises in the second period of competition. In this case, we assume that the bank will realize it if there is just one firm left in the market.

A4: Both firms and both banks are risk neutral.

A5: Production technology: As specified above, firms incur fixed production costs and invest in research expenditures in each period. The level of marginal costs depends on the success of their innovation activities.

A6: Consumer demand is specified as in the dynamic circular city-model above.

A7: The terms of contract for firm j are identical to those of firm i: The amount of loan extended in each period is I_t. The contract is based on reported profits and specifies repayments as well as refinancing probabilities, which are denoted by γ^D, γ^S, γ^A for firm j. Both financial contracts are observable.

The optimal financial contract

The expected value of firm j under external debt financing equals:

$$\begin{aligned}V_j =\ & \theta_i\theta_j[\Pi^S - R^S + \gamma^S(\overline{\Pi}_S^2 - g(\mu^S) - R^2)] \\ & + (1-\theta_i)(1-\theta_j)[\Pi^S - R^S + \gamma^S(\overline{\Pi}_S^2 - g(\mu^S) - R^2)] \\ & + \theta_j(1-\theta_i)[\Pi^A - R^A + 1\cdot(\Pi^{Mono} - g(\mu^{Mono}) - R^{Mono})] \\ & + \theta_i(1-\theta_j)[\Pi^D - R^D + 0] - g(\theta_j).\end{aligned} \quad (3.55)$$

If firm j is the only innovating firm in the first period, its refinancing probability equals $\gamma^A = 1$. If firm j fails to innovate in the first period while firm i is successful, firm j has to leave the market because the refinancing probability is zero, $\gamma^D = 0$.

The bank of firm j maximizes over expected repayments in both periods under *(i)* the limited liability conditions, *(ii)* the incentive compatibility restrictions and *(iii)* the optimal R&D activities of firm j:

$$\max\ W_j = \theta_i \theta_j R^S + \theta_i (1-\theta_j) R^D + (1-\theta_i) \theta_j R^A + (1-\theta_i)(1-\theta_j) R^S - I_1$$
$$+ \left[\gamma^S \theta_i \theta_j + \gamma^D \theta_i (1-\theta_j) + \gamma^S (1-\theta_i)(1-\theta_j) \right] (R^2 - I_2) \quad (3.56)$$
$$+ \left[\gamma^A (1-\theta_i) \theta_j \right] (R^{Mono} - I_2).$$

subject to the limited liability constraints

$$\Pi^D - g(\theta_j) \geq R^D, \quad \Pi^{DD,D,S} - g(\mu_j^D) \geq R^2,$$
$$\Pi^S - g(\theta_B) \geq R^S, \quad \Pi^{D,S,A} - g(\mu_j^S) \geq R^2, \quad (3.57)$$
$$\Pi^A - g(\theta_j) \geq R^A, \quad \Pi^{Mono} - g(\mu^{Mono}) \geq R^{Mono};$$

subject to the condition for optimal R&D expenditures that we derive by maximizing firm j's value (3.55) with respect to the innovation probability:

$$g'(\theta_j) = \theta_i [\Pi^S - R^S + \gamma^S (\overline{\Pi}_S^2 - g(\mu^S) - R^2)$$
$$- \theta_i (\Pi^D - R^D + 0]$$
$$+ (1-\theta_i)[\Pi^A - R^A + \Pi^{Mono} - g(\mu^{Mono}) - R^{Mono}] \quad (3.58)$$
$$- (1-\theta_i)[\Pi^S - R^S + \gamma^S (\overline{\Pi}_S^2 - g(\mu^S) - R^2)],$$

and the two binding incentive compatibility constraints:

$$\Pi^A - R^A + (\Pi^{Mono} - g(\mu) - R^{Mono}) \geq \Pi^A - R^S + \gamma^S (\Pi^{Mono} - g(\mu) - R^{Mono}), \quad (3.59)$$
$$\Pi^S - R^S + \gamma^S (\overline{\Pi}_S^2 - g(\mu^S) - R^2) \geq \Pi^S - R^D + \gamma^D (\overline{\Pi}_S^2 - g(\mu^S) - R^2). \quad (3.60)$$

Solution

The solutions to this optimal contract problem is derived as follows:

Again we know that if firm j is the only one whose innovation project fails in the first period, the respective profits will be insufficient to cover the loan repayment. Thus, the refinancing probability γ^D will be zero. In all other scenarios, the refinancing probabilities will be positive, but may be less than one. Because of the moral hazard problem, the repayments of the second period amount only to the smallest profit possible, Π^D-$g(\mu^S)$.

If we have a monopoly in the second-period, a bank will refinance its firm with probability $\gamma^A=1$, because the loan will be paid back with certainty. Since the bank has all the bargaining power, it wants to extract as much of the firm's monopoly profits as possible. But because of the competitiveness in the banking sector, the bank is unable to extract repayments higher than the amount of loan extended, I_2: If the bank asked for repayments R^{Mono} higher than I_2, another bank

3.3 Credit financing and product market competition

would offer an one-period credit contract to firm j for slightly less. Therefore, repayments in the case of second-period monopoly R^{Mono} will equal I_t.

By inserting these results into the incentive compatibility constraints and by taking the limited liability conditions into account, we derive the values for the refinancing probabilities and the repayments in the first and second periods. Again we have to distinguish between whether the innovation success probability μ^S is high or low.

Case 3 ($\mu^S < 0{,}5$): $\quad \gamma^D = 0$ and $R^D = \Pi^D - g(\theta_j)$

$\qquad\qquad\qquad\qquad \gamma^S = \gamma^A = 1$ and $R^S = R^A = \overline{\Pi}_S^2 - g(\theta_j)$

$\qquad\qquad\qquad\qquad R^2 = \Pi^D - g(\mu^S)$ and $R^{Mono} = I_2$

If the innovation success probability μ^S is less than fifty percent, (which depends on the cost parameter and on the level of the expected profits), the bank will grant second-period financing for firm j with probability $\gamma^S = \gamma^A = 1$.

Case 4 ($\mu^S \geq 0{,}5$): $\quad \gamma^D = 0$ and $R^D = \Pi^D - g(\theta_j)$

$\qquad\qquad\qquad\qquad \gamma^S = \dfrac{\Pi^S - \Pi^D}{\overline{\Pi}_S^2 - \Pi^D} < 1$ and $R^S = \Pi^S - g(\theta_j)$

$\qquad\qquad\qquad\qquad \gamma^A = 1$ and $R^A = (1 - \gamma^S)(\Pi^{Mono} - I_2) + R^S$

$\qquad\qquad\qquad\qquad R^2 = \Pi^D - g(\mu^S)$ and $R^{Mono} = I_2$

In this case the innovation probability μ^S of the symmetric market situation in period two is less than fifty percent, e.g. because R&D costs are very high, the refinancing probability γ^S will be smaller than one, and the first-period repayments R_j^A and R_j^S will be different.

By inserting these results into the condition for the optimal R&D expenditures (3.58), we see that the size of the monopoly profits has a negative impact on the slope, but a positive impact on the intercept of the reaction function of firm j:

$$g'(\theta_j) = \theta_i \cdot \underbrace{[\ldots - \Pi_i^{Mono} \ldots]}_{<0} + \underbrace{[\ldots + \Pi_j^{Mono} \ldots]}_{>0} \qquad (3.61)$$

If we do the same calculus for firm i, we obtain the intersection of the R&D reaction functions the optimal R&D levels when both firms are credit-financed.

$$\text{Case 3: } \theta_j^{d.f.} = \dfrac{\Pi^A + (\Pi^{Mono} - I_2) - \Pi^S + \Pi^D - \overline{\Pi}_S^2}{2\rho + \Pi^A + (\Pi^{Mono} - I_2) - 2 \cdot \Pi^S + 2 \cdot \Pi^D - \overline{\Pi}_S^2} = \theta_i^{d.f.} \qquad (3.62)$$

Case 4: $\theta_j^{d.f.} = \dfrac{\Pi^A + \gamma^S(\Pi^{Mono} - I_2) - 2\cdot\Pi^S + \Pi^D}{2r + \Pi^A + \gamma^S(\Pi^{Mono} - I_2) - 3\cdot\Pi^S + 2\cdot\Pi^D} = \theta_i^{d.f.}$. (3.63)

Proposition 3.10 (R&D activities if both firms are debt-financed)

If both firm i and firm j need external debt to finance the fixed production costs, their levels of optimal R&D activities in the first period will decrease. Compared to the unrestricted case we obtain:

$$\theta^{df.df} < \theta^{LT}.$$

As a consequence of the reduced R&D activities, product prices will increase.

Graphically, we see that in contrast to Figures 3.8 and 3.9 the reaction functions of the two credit-financed firms are symmetric again. In case the refinancing probabilities equal $\beta^S = \gamma^S = 1$, we obtain:

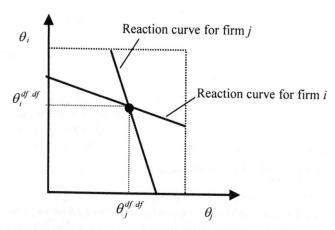

Figure 3.10: R&D reaction functions if both firms are loan-financed and refinancing probabilities equal $\beta^S = \gamma^S = 1$ (Case 3).

3.3 Credit financing and product market competition

In case the refinancing probabilities β^S, γ^S are smaller than one, we obtain the following shift of the first-period R&D reaction functions:

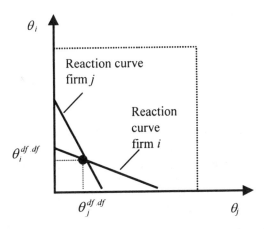

Figure 3.11: R&D reaction functions if both firms are loan-financed and refinancing probabilities are $\beta^S < 1$, $\gamma^S < 1$ (Case 4)

Impact of the two debt contracts on innovation, pricing, and competition

If both firms need external financing, R&D activities will definitely decrease. The financial contracts between the firms and their house bank imply that large parts of the innovation gains are transferred to the banks. So the firms have little incentives to engage in R&D activities.

Because of the decrease in R&D activities and the increase of the likelihood of a monopoly, product prices in the first and second period will be higher than in the case of self-financed firms.

If both firms are financed with external debt contracts, industry concentration will rise. Since product prices in this case are higher, overall welfare will decline.

Summary

The main point of the analysis is that external debt financing alters the innovative activities of both firms. Moreover, the optimal loan contract provides insufficient finance for the borrowing firm: Although expected net profits are positive in both periods, the firms in need of external finance will only obtain further financing if the actual first-period profits are not too low. The financial restriction leads to product market inefficiency. This inefficiency cannot be overcome since there is no scope for renegotiation: A bank will never agree to further finance a firm that has fallen behind in the innovation game of the first period because of the moral-hazard problem in the second period.

3.3.2.5 Predation

Theoretical findings

In the second part of their (1990) paper, Bolton and Scharfstein already address the question whether a long-term debt contract designed as above will induce a self-financed competitor to prey upon its credit-financed rival. The idea is that when the contract specifies a refinancing probability of less than one, the self-financed rival may attempt to increase the bankruptcy probability (i.e. reduce the success probability) of the credit-financed firm. This could be done via aggressive advertising, smear campaigns, or price dumping. If the predatory activities are successful, the bank will deny to refinance its firm in the second period, the leveraged firm will have to leave the market, and the self-financed rival will subsequently enjoy monopoly profits. [12]

In our innovation and competition framework, we investigate whether the self-financed firm j (cf. 3.3.2.3.) could prey upon its leveraged rival i. Since the terms of contract specify refinancing probabilities of less than one, firm j will engage in predation if additional profits are higher than the associated predation costs K^P:

$$(\theta_i - \theta_i^P)\theta_j[\Pi^A + \Pi^{Mono}] - (\theta_i - \theta_i^P)(1-\theta_j)[\Pi^D + E(\Pi_D^2)] > K^P. \quad (3.64)$$

Predation reduces firm i's success probability from θ_i to θ_i^P. At the same time, the likelihood for firm j to obtain a cost-advantage and to realize monopoly profits in the second period will increase and the probability to encounter a cost-disadvantage is reduced. Thus, if these additional expected profits exceed K^P, firm j will choose to prey.

The approach stands in the tradition of the long purse story (Telser (1966) in Tirole, 1988, 377): The firm with better access to liquid funds can survive longer in predatory product market competition.

---

[12] Snyder (1996) extends the Bolton and Scharfstein framework to contract renegotiation. One of the firms is financed by a Bolton-Scharfstein contract. The other, self-financed firm preys and attempts to reduce the refinancing probabilities of its rival, in order to induce the firm to exit the market. The bank and its firm therefore agree to renegotiate the financial contract in order to prevent predation. The approach of Snyder, however, does not model strategic interaction of firms in the product market nor does it take technological competition into account.

Empirical evidence

These findings are confirmed by empirical evidence, e.g. from the computer industry. Lerner (1995) investigates the disk drive industry and finds out that firms with sufficient internal funds did drive poorly capitalized rivals out of the market.

The disk drive industry is a good example for spatial competition in the form of our circular city-model. The attributes of the disk drives such as capacity (in megabytes), density (in bits per square inch), and access time (in milliseconds) can be characterized precisely. Among these, density is the most critical technical dimension. Product variants can, thus, be easily identified. As a consequence, the extent of price-undercutting between nearby product variants can be accurately measured.

While at the beginning of the 1980s most firms in the disk drive industry could self-finance their research expenditures and investments, profit margins declined substantially in 1984, and financing became much more difficult. Cash-rich manufacturers started to cut prices aggressively and sold similar drives 20 percent lower than those of their financially restricted competitors. Thinly capitalized firms, in turn, started losing market shares and finally had to withdraw from the market. Lerner's study thus supports the findings of our predation model.

3.3.3 Discussion

We presented a model of multi-stage innovation and product market competition in which two firms operate with or without financial constraints. Innovations are induced by firm-specific R&D activities. A successful innovation project reduces marginal production costs and enables the firm to charge a lower product price. The ensuing price constellation determines the market position and, hence, the profit situation of the firm.

If a firm is financially constrained, it applies for a bank loan in order to cover the fixed production costs. As for the information structure, we assumed that the bank is unable to observe the results of the innovation projects and, therefore, has no information about the actual profit situation of the firm.[13] The bank designs an optimal, incentive-compatible loan contract in which repayments and the chance to obtain follow-up financing are based on reported profits. Repayments in the first period exceed the amount of credit extended, whereas second-period repayments are restricted to the minimum profit level because of the moral hazard problem.

As we have seen, investment financing via these two-period, incentive-compatible credit contracts has the following impact on competition in oligopoly:

---

[13] An equivalent assumption is that firms are able to embezzle money before the bank can observe the realized profits.

If only one of the firms, e.g. firm i, needs external financing, the R&D activities of this firm decrease because large parts from the innovation gains are being transferred to the bank via the debt contract. The self-financed rival j, on the other hand, dramatically increases its R&D activities because a successful innovation in the first period gives her a cost-advantage and enables her to become a monopolist in the second period. As a consequence, firm i's first-period prices are higher and firm j's prices are lower than under self-financing. In the second period, however, firm j may be able to charge the monopoly price: If the debt-financed firm fails to innovate and has to leave the market, the self-financed firm has a chance to establish itself as monopolist.

If both firms need external financing, innovation activities in the industry will definitely decline. Product prices in the first and second periods are, thus, higher than under self-financing. Price competition between two debt-financed firms is, therefore, less intense. Moreover, the likelihood that one of the firms doesn't obtain refinancing increases. This implies that the rival becomes a monopolist and induces industry concentration to rise.

Finally, if the terms of the financial contract specify a refinancing probability smaller than one, a self-financed firm may decide to prey upon its debt-financed rival. Predation is profitable as long as additional profits from becoming a monopolist are greater than predation costs. Successful predation causes R&D activities to decline, product prices to rise, and the likelihood of monopoly to increase.

Consequently, market efficiency in the product market is reduced: The firms' need to obtain external financing implies that there is too little R&D investment and too little innovation. Thus, the intensity of competition declines.

How could the inefficiency be mitigated? Suppose that bargaining power in the financial contracting is not on the side of the banks but on the side of the firms. Empirically, this could be the case if financing was provided by federal funds, e.g. via the Reconstruction Loan Corporation (Kreditanstalt für Wiederaufbau) in Germany or via federal promotion programs for technology-based firms. In the present model, this alters the design of the optimal contract similar to Stadler (1997): The new objective function is given by the maximization problem of the firm subject to the zero-profit condition for the bank. As a consequence, the refinancing probabilities increase. Especially the refinancing probability for a firm which fails to innovate while the competitor is successful is now greater than zero. As a result, market efficiency increases.

3.4 Conclusion

In the present chapter, we first showed how problems of asymmetric information affected the borrower-lender-relationship between a firm and its bank: In case of moral hazard, a bank engages in costly state verification in order to prevent the entrepreneur from strategically defaulting on the repayment obligation. Since the monitoring activities are costly, the bank and the firm may agree to renegotiate the initial debt contract. In a dynamic setting, the moral hazard problem is reduced. Moreover, as Bolton and Scharfstein show, long-term debt contracts help to circumvent the moral hazard problem if costly state verification is impossible. The incentive-compatible debt contract simply has to specify smaller refinancing probabilities for low reported profits.

In the second part of the chapter, we combine financial debt contracts in the design of Bolton and Scharfstein with product market competition. We consider two well-established firms that engage in dynamic price competition with heterogeneous products. Firms are competing for market shares and engage in cost-reducing R&D activities. Our main results are that firms which have to rely on debt-financing, invest less in price-reduction and less in the acquisition of market shares. Moreover, their R&D intensity declines. On the other side, incentives for self-financed rivals to engage in predation increase, which forces under-capitalized firms to exit.

So far, we have focused on competition between two well-established firms. In the next chapter, we turn to an alternative scenario: We consider a young firm that attempts to enter a market where it faces price competition with an incumbent firm. Since bank loan financing is generally not available for a young start-up firm, the firm has to find other financing alternatives. In the next chapter, we first point out that venture capital financing is one of the most important financing source for fast-growing start-up firms. Then, we analyze the financial contracts between the young firm and the venture capital investor. After this, we investigate how the financial contract affects market entry and the subsequent price competition game. We show, that a self-financed incumbent has incentives to engage in entry deterrence and predation.

3.5　Appendix

R&D reaction functions in the second period (Subsection 3.3.2.2)

The R&D reaction functions in the symmetric cases have the following form:

$$2\rho\mu_i^S = \underbrace{\left[-\frac{1}{9}\frac{(\Delta c)^2}{T}\right]}_{\zeta} \cdot \mu_j^S + \underbrace{\left[\frac{(\Delta c)}{3} + \frac{(\Delta c)^2}{9}\frac{1}{2T}\right]}_{\iota} \tag{A1}$$

$$\mu_i^{S*} = \frac{\iota \cdot (2\rho + \zeta)}{(2\rho - \zeta)(2\rho + \zeta)} = \mu_j^{S*} \tag{A2}$$

In the asymmetric cases, we have for the firm with the technological advantage:

$$2\rho\mu^A = \underbrace{\left[-\frac{1}{9}\frac{(\Delta c)^2}{T}\right]}_{\zeta} \cdot \mu^D + \underbrace{\left[\frac{(\Delta c)}{3} + \frac{(\Delta c)^2}{3}\frac{1}{2T}\right]}_{\vartheta} \tag{A3}$$

$$\Leftrightarrow \mu^{A*} = \frac{\zeta\varpi + \vartheta 2\rho}{(2\rho - \zeta)(2\rho + \zeta)} \tag{A4}$$

and for the firm with the technological disadvantage:

$$2\rho\mu^D = \underbrace{\left[-\frac{1}{9}\frac{(\Delta c)^2}{T}\right]}_{\zeta} \cdot \mu^A + \underbrace{\left[\frac{(\Delta c)}{3} - \frac{(\Delta c)^2}{9}\frac{1}{2T}\right]}_{\varpi} \tag{A5}$$

$$\Leftrightarrow \mu^{D*} = \frac{\zeta\vartheta + \varpi 2\rho}{(2\rho - \zeta)(2\rho + \zeta)} \tag{A6}$$

We see that the slope of all three second-period R&D reaction functions is negative and of the same size: Their steepness is determined by the (quadratic) cost-advantage obtained through innovation, which is normalized to consumer disutilities, i.e. transportation costs. If transportation costs are high, cost-reducing innovation becomes less important to the firm, because consumers are loyal to their previously chosen product variant. The intercept of the R&D reaction functions is highest for the firm that starts with a cost-advantage, medium if firms have symmetric market positions, and shifts toward the origin if a firm starts with a cost-disadvantage in the second period.

3.5 Appendix

Comparative statics for the optimal R&D levels μ^{S*}, μ^{A*} and μ^{D*} yield the following results:

The denominators $(2\rho+\zeta)(2\rho-\zeta)$ are identical and greater than zero for $\rho > (\Delta c)^2/18T$; which is fulfilled for most parameter constellations. As for the numerators, we see that $9 > \iota > \varpi$. Therefore, by rearranging terms, we can derive that the cost-leader will always invest more in second-period R&D activities than the cost-follower, i.e. $\mu^A > \mu^D$. However, second-period innovation activities in the symmetric case are even higher than those of the cost-leader, if the following inequality holds: $\iota(2\rho+\zeta) > \zeta\varpi + 2\rho\vartheta$, i.e.

$$\left(\frac{(\Delta c)}{3} + \frac{(\Delta c)^2}{9}\frac{1}{2T}\right)\left[2\rho + \frac{1}{9}\frac{(\Delta c)^2}{T}\right] - \left(-\frac{1}{9}\frac{(\Delta c)^2}{T}\right)\left(\frac{(\Delta c)}{3} - \frac{(\Delta c)^2}{9}\frac{1}{2T}\right) \quad (A7)$$
$$-\left[2\rho\left(\frac{(\Delta c)}{3} + \frac{(\Delta c)^2}{3}\frac{1}{2T}\right)\right] > 0.$$

Simplification leads to:

$$\frac{2}{9}\frac{(\Delta c)^2}{T}\left[\frac{c}{3} - \rho\right] > 0. \quad (A8)$$

Since the first term is always positive, we derive that if $\rho < (\Delta c)/3$, innovative activities are highest when firms have equal market positions. Q.E.D.

4 Venture capital financing and strategic competition

> *Equity financing is modeled when cash flows and asset values are not verifiable. Investors have enforceable property rights to the firm's assets, but cannot prevent insiders (managers or entrepreneurs) from capturing cash flow.*
>
> Stewart Myers (2000)

In this chapter we first point out the financing alternatives for small and medium sized firms. Our analysis follows the financial growth cycle (see Berger and Udell 1998), which indicates the type of financing (debt or equity) a young firm receives at various stages of its life. The financial growth cycle states that newly founded firms are typically not eligible for bank loan financing due to the lack of collateral and reputation.[1] Instead, a young firm has to rely on private equity or venture capital financing. We assume that the young firm has an innovative idea and needs external funding to realize a product innovation.

Equity financing has long been a neglected area of research, some scholars called it even uninteresting. Since the 1990s, however, public attention has focused on entrepreneurial enterprises which are considered as the engine of economic growth. New ventures create employment and spur innovation. Product innovation and productivity gains - particularly in the high-tech, information and life science areas – are vitally dependent on a flourishing entrepreneurial sector. Moreover, with the creation of the Neuer Markt (the German stock exchange for young high-tech firms) on March 10th, 1997, in Frankfurt, financial investors in these start-up firms now have an institutionalized exit channel to cash out their financial engagement. In the last years, venture capital financing, therefore, has become a field of special interest.

---

[1] This stands in contrast to the Poitevin (1989) model (see 2.2.3), which states that young firms rely on debt financing.

The chapter is organized as follows: In the Introduction (4.1) we describe the financial growth cycle of a young firm. After this, we point out the special characteristics of venture capital financing, which is an important source of funding for growth-oriented start-up firms. In section (4.2) we analyze the bilateral contracts between a young firm and a venture capital company. The main question of the models that we examine below is how control and ownership rights are distributed in the contractual arrangements. To our great surprise, none of these papers takes the competitive environment of the start-up firm into account. Many start-up ventures fail, however, because strategic reactions from competitors have not been taken into account. In section (4.3) we, therefore, analyze venture capital financing of an innovation project in the framework of industrial organization: We formalize the idea of market entry and competition between the start-up entrepreneur and an incumbent of the same industry. If the venture capital-backed firm successfully enters the market, it will face dynamic price competition with the incumbent firm. As a next step, we analyze how the incumbent's strategic reactions will influence the financial contracting between the young firm and its equity investor. We show that if the incumbent engages in strategic competition or predation, venture capital companies will prefer to finance later-stage investments. In this case it will become much harder for the young firm to obtain venture capital financing at all. Section (4.4) summarizes our results.

4.1 Introduction

In the introduction we describe the sources of small business finance and illustrate the stylized facts of venture capital financing.

4.1.1 The financial growth cycle

The financial growth cycle paradigm states that different capital structures are optimal at different stages of a growing firm. The financial growth cycle describes the various sources of small business finance and how the type of financing varies with firm size, age, and information availability. As the small business grows, financial needs and options change. Figure 4.1 shows this in a stylized fashion in which firms lie on a size/age/information continuum.

4 Venture capital financing and product market competition

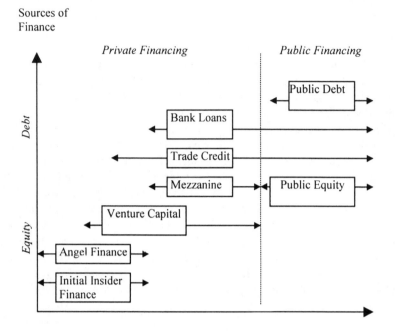

Figure 4.1: The financial growth cycle. Own presentation. See also Sidler (1997), Schefczyck (2000), Berger and Udell (1998), Bundesverband deutscher Kapitalbeteiligungsgesellschaften ("BVK") (2001).

At the beginning, firms must rely on initial insider financing provided by the start-up team, their family and friends. After the firm's foundation, business angels may be willing to make private equity investments and provide expertise and advise. As the firm begins small scale production and develops a formal business plan, it gets access to intermediate finance on the equity side, i.e. venture capital financing. Venture capital is used to finance either early-stage development, or later-stage market expansion after the new product has already been successfully test-marketed. Mezzanine funds represent high-yield loans with an "equity kicker". Mezzanine financing is typically used to bridge the equity gap between venture capital financing and the initial public offering. An IPO represents the first access to public capital markets, but only for the top segment of young, growing firms. Bank loans are typically not available to small businesses until they break-even and begin to accumulate substantial assets that can be pledged as collateral. Loans are a much cheaper form of financial instrument because the investors are rewarded only for the downside risk, but do not participate in the firm's upside potential. Trade credits are available from a rather early stage on, however, they represent an expensive form of short-term loans. Only very large firms with sufficient collateral and a well-known track record may issue shares or even commercial bonds on public capital markets.

The sequencing of funding over the growth cycle of a firm can be viewed in the context of the modern information economics: The fact that high-risk, high-growth firms often obtain venture capital finance before they obtain external debt finance suggests that moral hazard and adverse selection problems may be particularly acute for these firms: The information structure is such that the quality of the innovation project is initially uncertain and that the entrepreneur's behavior is unpredictable, since she has no track record for being a good manager or a reliable debtor. The "informational opaqueness" (Berger and Udell 1998) may, on the one hand, induce the entrepreneur to moral hazard: She may spend insufficient effort on the innovation project or may divert the funds to her private ends. On the other hand, the informational opaqueness may result in a lemon's problem (Akerlof 1970) in that the pool of projects seeking external equity finance consists of rather low quality projects. Venture capital companies as financial intermediaries, therefore, specialize in screening and monitoring activities: Before funds are provided, venture capital companies perform due diligence studies to evaluate the project's quality. Moreover, they actively monitor the entrepreneur's behavior during the investment's horizon of the financial engagement.

Information economics has also produced the notion of a financial pecking order (Myers 1984, Myers and Majluf 1984). The pecking order hypothesis suggests that when capital markets are imperfect due to informational asymmetries, a firm should first use its internal funds, then apply for bank loans, and only as a last resort issue new shares to finance a new investment opportunity. The first part of the ranking is caused by a problem of moral hazard, which requires costly state verification (Townsend 1979, Diamond 1984) and suggests the optimality of debt contracts after internal self-financing has been exhausted. The second part is due to a problem of adverse selection (Myers 1984, Myers and Majluf 1984, Nachman and Noe, 1994), which states that only firms whose stocks are overpriced will issue new shares on the public capital market.

How does this financial pecking order hypothesis stand in relation to the financial growth cycle? First of all, the pecking order hypothesis reflects the financing alternatives for large, well established companies that are listed on the stock exchange. These companies have already attained a later stage of the financial growth cycle, and can, therefore, choose between external debt and external equity. Small businesses, however, do not have access to these financial instruments due to high failure risk and severe information problems. Moral hazard can make debt contracts quite problematic, especially when the amount of external financing is large relative to the amount of internal funds. This suggests that private equity finance may be particularly important when this condition holds and moral hazard problems are acute. We conclude that the pecking order hypothesis does not cover the entire growth horizon of a firm and neglects the questions of small business finance.

Thus, before we analyze the financial contracts between venture capitalists and young start-up firms, we want to highlight the special characteristics of small business finance.

4.1.2 Characteristics of venture capital financing

Venture capital financing represents the main source of funding for fast-growing firms in the early and expansion stages. What are the stylized facts of venture capital financing? The following list summarizes the main features:

1. Venture capital backed innovation projects are very risky, yet they offer high growth and high profit opportunities: Successful venture capital engagements yield an average return on investment of about 30 percent per year. On the other hand, about 20% to 30% of the projects will result in a total loss (BVK statistics 2000, 62p., Gompers and Lerner 1999c).

2. Financial support is given in the form of equity financing: This implies that the venture capital investor is legally responsible and participates in gains and losses.

3. Funding is provided in accordance with the project's development: Initially, the venture capital company finances only a part of the total investment. The next tranche of capital is infused only after pre-specified milestones are attained.

4. Screening activities: Before the financial contract is signed, venture capital companies undertake strenuous efforts in sorting and screening the investment projects: Only three to five projects out of 100 business plan applications will typically receive funding. The project selection consists of a multi-stage decision process:

1.	VC company obtains business plan (directly from the start-up company, from a bank, or at an information exchange forum)	
2.	Pre-selection (VC company checks product, market environment, income forecasts, turnover figures)	100 %
3.	Main selection (personal contact with founders, discussion of management strategies, technical, financial and legal due diligence)	25 %
4.	Equity stake negotiations (valuation of the start-up company, investment level, financing instruments, information and control rights)	8 %
5.	Conclusion of contract, implementation, consulting activities from part of the venture capital company	3 %

Table 4.1: Multi-stage selection process before venture capital contracts are signed. See also Schefczyk (2000, 39pp.), Sidler (1997, 82pp.), BVK statistics (2001, 2000).

Table 4.1 illustrates that there is a vast number of innovative ideas, yet only a tiny fraction will turn out to be marketable or profitable. To screen potential projects, venture capitalists have to assess the market segment, the project's technology, the firm's management qualification, and the strength of existing competitors. Only when these criteria are approved, the venture capital company enters into equity stake negotiations: The young firm determines its financing needs (e.g. EURO 2,6 Mio. for the first investment period), then the venture capital company calculates the young firm's value (e.g. EURO 13 Mio.) and the percentage of shares it want to receive in exchange for the financial engagement (20 %). If both parties agree, the financial contract is signed and the venture funds are provided.

5. Informational opaqueness: Although the relationship between entrepreneur and venture capital investor is pretty close, agency problems still arise due to limited information availability in start-up financing. Adverse selection is mitigated via the pre-contractual due diligence process. Moreover, venture capitalists can stop bad quality projects and refuse follow-up financing. In addition to this, moral hazard can still be a problem, especially if the entrepreneur, but not the investor is familiar with the technical details of the venture project: The entrepreneur may derive a private benefit from the scientific or technical success, whereas the venture capital company is more interested in cash flow, marketing, and sales figures.

6. In addition to the supply of funds venture capital companies provide advise and expertise. The start-up entrepreneur, though an expert in technology, is often a beginner in business administration. The venture capital company may, thus, assist in elaborating strategies, recruiting key-employees and building up contacts to clients, banks, and to strategic investors.

7. The investment is long-term, usually three to five years. Venture capital is typically provided from a closed-end fund which lasts as long as ten years. If the project is good, the venture capital company wants to participate in subsequent financing rounds, because it is already familiar with the project's estimated quality.

8. Distribution of information- and control rights: Venture capital companies receive monthly data on sales revenues, the balance sheet, and profit and loss accounts. However, due to the uncertain environment and the limited experience of the start-up entrepreneurs venture capital companies usually have the right to intervene in the project. Even if venture capital companies own less than 50 percent of the voting rights, they still can take over control or have veto-power (Schefzcyk 2000, 42).

9. Divestment possibilities: The most common exit channels are an initial public offering (IPO) or a trade sale in case of success, and a buyback or project liquidation in case of failure. Figure 4.2 shows the distribution of exit channels.

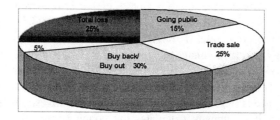

Figure 4.2: Exit channels for venture capital investors. Own presentation. See also BVK statistics (2001, 1999), Schefczyk (2000, 43-45).

Going public (15%) is still the exit channel most start-up firms dream about. An initial public offering, however, is only possible for projects with a very strong upside potential. Problems associated with an admission to the stock exchange are stricter disclosure requirements, the costs of establishing the company's legal organs (general meeting of shareholders, supervisory board), the cost of underpricing, and the possibility of hostile takeover bids.

A trade sale (25%) is the second most attractive exit channel. The start-up firm is sold to a company of the same industry, when the established firm's strategic benefit exceeds the value of start-up firm. The transaction itself is less costly and less time-consuming than an initial public offering, because competitors, i.e. potential buyers, are usually well known.

In case of a buyback or a management buyout (30%), the original founders or an external management team repurchase the venture capital company's equity stake. A problem are the limited financial resources of the founders. Buyouts are often highly leveraged transactions, in which around 80 percent of the cash flow is used for debt services. This restricts the firm's future operating alternatives. Buybacks represent a rather modest exit channel. If these projects are kept for a long time in the portfolio, they are called "living deads", because they generate neither big gains nor big losses.

Of minor importance are secondary purchases (5%). This exit channel implies that the venture capital company sells her equity stake to another financial intermediary.

The remaining fraction of projects (25%) are failures. Venture capitalists have to liquidate or abandon them.

10. Financial instruments: Preferred shares and convertible securities are most common. Preferred shares have a priority right to dividend payments. Convertible securities represent shares that can be converted into senior debt if the young firm does not attain its milestones and the projects fails.

To summarize: What are the main features of venture capital financing, especially in comparison to "standard" bank loan financing? First of all, venture capital companies specialize in identifying high-growth projects. As equity investors they participate not only in the downside risk, but also in the upside potential of innovation projects. Venture capital companies engage in early-stage and expansion–stage financing. In the start-up phase, however, the cash-flow of a firm is negative. Thus, loan financing is impossible: Credits are not supplied, because projects have high failure risk and the entrepreneur cannot offer collateral. Interest payments would be exorbitantly high if the risk premium was included. Venture capital companies, therefore, choose adequate financing tools (staged financing, design of special financial instruments) in order to control or discipline the management. In addition to that, they provide consulting services. All this makes venture capital financing the most important source of funding in early stage.

4.2 Venture capital financing – the individual firm's perspective

Shareholders are stupid and impertinent – stupid because they give their money to somebody else without any effective control over what this person is doing with it, and impertinent because they ask for a dividend as a reward for their stupidity.
Carl Fürstenberg (1850-1933), banker

Most theoretical models on venture capital financing have focused on the bilateral relationship between a young start-up firm and the venture capital company so far. This financial relationship is analyzed in the framework of the contract theory. The existing theoretical literature on venture capital financing can be classified into two main groups: The first group investigates the distribution of control rights in venture capital contracting. The second group is interested in how the financial contract solves the distribution of property rights. In this subsection 4.2, we give a short overview on these existing contracting models, before we turn to our analysis of the industry level.

4.2.1 Allocation of control rights

Regarding the allocation of control rights in venture capital contracting, the central question is who should manage the project, the entrepreneur or the venture capitalist? The models assume that the entrepreneur is an expert in technology, but a beginner in business administration. Thus, when the entrepreneur's skill level turns out to be insufficient, a conflict of interest arises: The venture capital company's wishes to replace the entrepreneur by a professional management, but the entrepreneur derives a personal benefit from running the innovation project. Papers from this area include the work of Amit, Glosten, and Muller (1990), Chan, Siegel, and Thakor (1990), Berglöf (1994), Marx (1998), and Hellmann (1998). We exemplify the control rights approach by presenting the model of Chan, Siegel, and Thakor (1990). Their model includes several important features of venture capital contracting, such as an ex ante unknown skill parameter of the entrepreneur and the venture capitalist's right to intervene in the project. After this rather formal approach, we briefly describe the related ideas of Hellmann (1998) and Marx (1998).

4.2.1.1 Assignment of control rights

Chan, Siegel, and Thakor (1990) show that an entrepreneur must exhibit above-average management skills in order to retain control of the venture-capital-backed innovation project.

Assumptions

A1: Players: There is a wealth-constrained entrepreneur who has a creative project idea. A venture capital company that has sufficient funds to finance the project. The venture capital market is characterized by perfect competition.

A2: The time horizon is one period which consists of two stages: an investment stage and an action stage. *(i)* In the investment stage, the financial contract is signed and the amount I is invested. After this, the entrepreneur's skill level

γ_y (y = young firm) is revealed. Depending upon the realization of γ_y, the control of the young firm will be in the hands of either the entrepreneur or the venture capitalist. *(ii)* In the action stage, the party in control has to undertake effort e, which determines the project's success probability $\theta(e)$ and the respective failure probability $1-\theta(e)$. At the end of this stage, the terminal cash flow is realized and the entrepreneur and the venture capitalist share the returns according to the contract specifications.

A3: Information structure: There is pre-contract homogeneity in information and beliefs: Both parties share common priors about the entrepreneur's skill level $\gamma_y \in [\underline{\gamma}, \overline{\gamma}]$ and its probability distribution $F(\gamma_y)$. The venture capitalist's skill level is common knowledge and is normalized to $\gamma_{VC}=1$. The effort level undertaken by either party in control, e_y or e_{VC}, is unobservable and, therefore, not contractible.

A4: Risk preferences: The entrepreneur is assumed to be risk-averse. The utility she derives from engaging in the project is given by the additive-separable von Neumann-Morgenstern utility function $U_y(\Pi, e_y) = u(\Pi) - v(e_y)$, where the functions $u(.)$ and $v(.)$ are defined on $\Re^+$. The function $v(e_y)$ captures the entrepreneur's disutilities of effort. The function is increasing and concave, $v'(e_y)>0$, $v''(e_y)<0$, and satisfies the Inada-conditions $v'(0) \to \infty$ and $v'(\infty) \to 0$. If the venture capital company is in control of the project, the entrepreneur's disutility of effort is zero, $v(.)=0$. We define the inverse function $\psi(.) = u^{-1}(.)$, which is strictly increasing and strictly convex in $\Re^+$ in order to express the util payments in monetary payoff terms (see also Grossman and Hart 1983).

The venture capital company, on the other hand, is risk-neutral, because it has a large portfolio of diversified investment projects. If the venture capital company controls the project, it has the same disutility of effort function $v(e_{VC})$ as the entrepreneur. We define the same inverse function $\psi(.)$ to derive the pecuniary equivalent of the venture capitalist's disutility of effort.

A5: The production technology is given only in reduced form, i.e. it is completely described by the probability distribution of the terminal cash flow. The project outcome is dichotomous: The returns can be either high or low, $\Pi^H > \Pi^L > 0$. They depend linearly on the skill level of the party in control. Second, the success probability $\theta(e)$, which is determined by the effort level of the party in control, is strictly increasing, twice continuously differentiable and concave in e. The failure probability is given by $1-\theta(e)$. Thus, the project's expected gross profit equals $\theta(e)\gamma\Pi^H + [1-\theta(e)]\gamma\Pi^L$ with $\gamma = \{\gamma_{VC}, \gamma_y\}$ and $e = \{e_{VC}, e_y\}$.

116   4 Venture capital financing and product market competition

A6: Due to the reduced form of the project's returns, consumers' demand is not explicitly given.

A7: Financial contracting: The entrepreneur makes a take-it-or-leave-it offer to the venture capital company. The venture capital company competes to give the entrepreneur the highest expected utility, subject to the information constraint and the venture capitalist's participation constraint. Moreover, the contract must stipulate who is in control of production. Note, however, that the contract specifies ownership and control rights separately, i.e. it is not the party that owns the majority of equity which controls production, but the one that provides the best skill level for project realization. This is a common feature of venture capital contracting, i.e. even if the venture capitalist holds only a minority of equity, she typically has a veto right in major decision making.

A8: Distribution of control rights: At the beginning, the entrepreneur manages the project. If after the first stage the entrepreneur's skill level is revealed to be low, i.e. $\gamma_y \in [\underline{\gamma}, \hat{\gamma}]$, the venture capitalist obtains control to manage the project and to choose effort level e_{VC}. If the entrepreneur is relieved of productive control, she is paid a fixed amount independent of her demonstrated skill (to be shown below). If in the first stage the entrepreneur's skill level turns out to be high, $\gamma_y \in (\hat{\gamma}, \overline{\gamma}]$, control remains with the entrepreneur and both the venture capitalist and the entrepreneur receive risky payoffs. The indicator variable $M(\gamma)$ takes the value $M(\gamma)=1$ if the entrepreneur remains in control, and $M(\gamma)=0$ if the venture capitalist has second-stage control over the project.

The maximization problem

The state-contingent contract is given by $\{S^L \gamma \Pi^L, S^H \gamma \Pi^H, M(\gamma)\}$, where S^L, S^H represent the entrepreneur's shares of profits, γ denotes the skill level of either the entrepreneur or the venture capitalist, and $M(\gamma)$ indicates which party is in control of production. The financial contract solves the entrepreneur's maximization problem

$$\max_{S^H, S^L, M(\gamma)} U_y = \int_{\underline{\gamma}}^{\overline{\gamma}} \{\theta(e^*) \cdot u(S^H \gamma \Pi^H) + [1-\theta(e^*)] \cdot u(S^L \gamma \Pi^L) - v(e^*) M(\gamma)\} dF(\gamma_y).$$

(4.1)

Given the skill level γ_y, the entrepreneur maximizes the utility that she derives from the payoff shares in case of a high or a low project outcome, less the effort disutilities she must bear if she controls the project. In the above expression, e^* represents the optimal effort level for a given contract.

4.2 Venture capital financing – the individual firm's perspective

The maximization problem is subject to the state-contingent participation constraint of the venture capital company.[2] The venture capitalist's participation condition, if the entrepreneur is in control, is given by:

$$\int_{\hat{\gamma}}^{\bar{\gamma}}\left\{\theta(e_y)(1-S^H)\gamma_y\Pi^H + [1-\theta(e_y)](1-S^L)\gamma_y\Pi^L\right\}dF(\gamma_y) \geq I$$

for $M(\gamma)=1$. (4.2)

It says that for any given skill level, the venture capitalist's expected share of profits must be equal to or greater than the initial investment I. The critical level of the skill parameter $\hat{\gamma}$ above which the entrepreneur retains control is specified below.

If, on the other hand, the venture capital company is in control of the project, the skill parameter is normalized to unity, $\gamma_{VC}=1$. The participation constraint incorporates the effort cost for the venture capitalist and is given by:

$$\int_{\underline{\gamma}}^{\hat{\gamma}}\left\{\theta(e_{VC})(1-S^H)\Pi^H + [1-\theta(e_{VC})](1-S^L)\Pi^L\right\}dF(\gamma_y) - \psi(v(e_{VC})) \geq I$$

for $M(\gamma)=0$. (4.3)

Note here that the venture capitalist's disutility of effort function is transformed into the pecuniary equivalent by the function $\psi(v(e_{VC}))$. Since the optimal effort e^* depends on who controls production, we define $e_y=e^*$ as the optimal value if the entrepreneur is in control, and $e_{VC}=e^*$ if the venture capitalist is in control.

Second-stage control decision

The solution strategy to the problem above, i.e. the optimal contract, is required to be renegotiation proof (see Fudenberg and Tirole 1988), in the sense that once the skill level γ_y is revealed after the first stage, neither the entrepreneur nor the venture capitalist must have an incentive for mutually beneficial recontracting. This implies that the state-contingent contract must be ex-post Pareto-optimal. Pareto-optimality means that neither the entrepreneur nor the venture capitalist are able to increase their own utility without decreasing the other party's utility. We solve by backward induction and derive the "contract curve" for the venture capitalist and the entrepreneur for any given γ_y. We consider separately the case in which the entrepreneur is in control and that in which the venture capitalist is in control. We then compare both solutions to derive the optimal choice of control.

---

[2] Since the venture capital company is assumed to be risk-neutral, its payoffs need not be transformed by a utility function.

Case M(γ)=1 (The entrepreneur is in control)

We assume that the entrepreneur wants to attain at least an expected utility of $\overline{U}_y$. The venture capitalist then maximizes her expected payoff. Because the Grossman and Hart (1983) convexity of distribution function condition is supposed to hold here,

$$\frac{\partial^2 F(\Pi|e_y)}{\partial e_y^2} > 0, \qquad (4.4)$$

we can use the first-order approach[3] to characterize the entrepreneur's optimal choice of effort, e_y. Thus, for a given skill level γ_y and the young firm's reservation utility $\overline{U}_y$, the venture capitalist maximizes her expected utility by

$$W(\gamma_y, \overline{U}_y)\big|_{M(\gamma)=1} = \max_{(1-S^L),(1-S^H)} \theta(e_y)(1-S^H)\gamma_y\Pi^H + [1-\theta(e_y)](1-S^L)\gamma_y\Pi^L, \quad (4.5)$$

subject to the entrepreneur's reservation utility

$$\theta(e_y) \cdot u(S^H \gamma_y \Pi^H) + [1-\theta(e_y)] \cdot u(S^L \gamma_y \Pi^L) - v(e_y) \stackrel{!}{=} \overline{U}_y, \qquad (4.6)$$

and the first-order condition for the optimal choice of action,

$$\theta'(e_y) \cdot (\Pi^H - \Pi^L) = v'(e_y). \qquad (4.7)$$

Equation (4.7) implies that marginal expected profits must equal marginal disutilities of effort.

Case M(γ)=0 (The venture capital company is in control)

In this case, the entrepreneur is needed only because she "owns" the initial project idea. Thus, for a given reservation utility $\overline{U}_y$, the venture capitalist controls the project and maximizes her expected utility by

---

[3] The first-order approach is problematic because the second-order sufficiency condition cannot be interpreted in an economically plausible way (see Mirrlees 1973). Thus, the first-order approach is viable only if special assumptions about the distribution function are made: Grossman and Hart (1983) show that both the monotone likelihood ratio property and the convexity of distribution function condition must hold. Jewitt (1988) and Sinclair-Desgagne (1994) formulate more general conditions for which the first-order approach is fulfilled.

4.2 Venture capital financing – the individual firm's perspective

$$W(\gamma_{VC}, \overline{U}_y)\big|_{M(\gamma)=0} = \max_{(1-S^L),(1-S^H),e_{VC}} \theta(e_{VC})(1-S^H)\Pi^H \\ + [1-\theta(e_{VC})](1-S^L)\Pi^L - \psi(v(e_{VC})). \quad (4.8)$$

This maximization problem is subject to the entrepreneur's utility constraint

$$\theta(e_{VC})u(S^H\Pi^H) + [1-\theta(e_{VC})]u(S^L\Pi^L) \stackrel{!}{=} \overline{U}_y, \quad (4.9)$$

as well as to the first-order condition for the venture capitalist's optimal effort choice,

$$\theta'(e_{VC}) \cdot [(\Pi^H - \Pi^L) - S^H\Pi^H - S^L\Pi^L] = \psi[v(e_{VC})] \cdot v'(e_{VC}). \quad (4.10)$$

Equation (4.10) states that the optimal action $e^* = e_{VC}$ chosen by the venture capitalist is found where the expected share of marginal profits accruing to the venture capitalist equals the pecuniary equivalent of the marginal disutilities of effort.

Comparison

By comparing both cases $M(\gamma)=1$ and $M(\gamma)=0$, the venture capitalist's expected payoff in the second stage will be higher with the entrepreneur managing the project if and only if

$$W(\gamma_y, \overline{U}_y)\big|_{M(\gamma)=1} \geq W(\gamma_{VC}, \overline{U}_y)\big|_{M(\gamma)=0} \quad (4.11)$$

If condition (4.11) is violated, the venture capitalist's expected payoff is higher if she manages the project herself. Thus, the decision about who controls the project is given by

$$M^*(\gamma, \overline{U}_y) = \begin{cases} 1 & \text{if } W(\gamma_y, \overline{U}_y)\big|_{M(\gamma)=1} \geq W(\gamma_{VC}, \overline{U}_y)\big|_{M(\gamma)=0} \\ 0 & \text{if } W(\gamma_y, \overline{U}_y)\big|_{M(\gamma)=1} < W(\gamma_{VC}, \overline{U}_y)\big|_{M(\gamma)=0}. \end{cases} \quad (4.12)$$

To sum up, given the entrepreneur's reservation utility, renegotiation proofness completely specifies the venture capitalist's optimal financial contract in the second-stage, $\Gamma^2 = \{S^L(\gamma_y, \overline{U}_y), S^H(\gamma_y, \overline{U}_y), M^*(\gamma, \overline{U}_y)\}$, and the corresponding effort choices, e_{VC} and e_y.

First-stage utility maximization

In order to derive an overall solution to the program (4.1) to (4.3), we take the ex-post optimal second-stage contract Γ^2 as given and search for the optimal utility allocation for an entrepreneur of skill γ_y, regardless of who is in control of the project. This means that the entrepreneur maximizes

$$\max_{\bar{u}(\gamma)} \int_{\underline{\gamma}}^{\hat{\gamma}} \overline{U}_y(\gamma) dF(\gamma) + \int_{\hat{\gamma}}^{\bar{\gamma}} \overline{U}_y(\gamma) dF(\gamma), \tag{4.13}$$

subject to the participation constraint of the venture capital company

$$\int_{\underline{\gamma}}^{\hat{\gamma}} W(\gamma_{VC}, \overline{U}_y(\gamma))\big|_{M(\gamma)=0} dF(\gamma) + \int_{\hat{\gamma}}^{\bar{\gamma}} W(\gamma_y, \overline{U}_y(\gamma))\big|_{M(\gamma)=1} dF(\gamma) = I, \tag{4.14}$$

where $\hat{\gamma}$ indicates the critical skill value below which the regime of control switches from the entrepreneur to the venture capitalist. The critical value is determined by

$$\hat{\gamma} = \{\gamma \in [\underline{\gamma}, \bar{\gamma}] \mid W(\gamma_{VC}, \overline{U}_y(\gamma))\big|_{M(\gamma)=0} \stackrel{!}{=} W(\gamma_y, \overline{U}_y(\gamma))\big|_{M(\gamma)=1} \}. \tag{4.15}$$

The solution precludes bilateral incentives to renegotiate the contract once the skill level is revealed.

Solution

In case $\gamma_y < \hat{\gamma}$, the venture capitalist controls the project in the second stage, and the entrepreneur is paid a fixed amount regardless of the terminal cash flow and regardless of the entrepreneur's skill level γ_y:

$$u(S^L \Pi^L) = u(S^H \Pi^H) = \overline{U}_y. \tag{4.16}$$

Proposition 4.1 (Shift of control to the venture capitalist)

If the entrepreneur's skill level is below the critical value, $\gamma_y < \hat{\gamma}$, the venture capitalist takes control over the project, and the entrepreneur is paid a fixed amount which gives her the reservation utility in either state of nature.

Proposition 4.1 holds because the first-best risk sharing arrangement involves the risk-neutral venture capital company completely insuring the risk-averse entrepreneur against cash flow randomness. Since the venture capitalist is controlling production here, the fixed payment to the entrepreneur achieves both optimal risk sharing and efficient solution of the moral hazard problem (because

4.2 Venture capital financing – the individual firm's perspective

there is no moral hazard problem for the entrepreneur). Moreover, the entrepreneur's compensation does not depend on her skill level. The reason behind this is that the allocations across different skill level realizations have to be Pareto-efficient.

By contrast, if the entrepreneur's skill level is above the critical value, $\gamma_y \geq \hat{\gamma}$, control remains with the entrepreneur. Chan, Siegel and Thakor show that the entrepreneur's skill level will now necessarily be greater than one, $\gamma_y > 1$. This is due to the fact that the entrepreneur's action choice is unobservable and that the entrepreneur is induced to undersupply effort $e_y < e^*$ (similar to Holmström 1983). Hence, in order to circumvent the moral hazard problem, it will be efficient to have the venture capitalist manage the project up to the critical value $\hat{\gamma} > 1$, even though the entrepreneur's skill level here already exceeds the venture capitalist's skill level by small values.

Proposition 4.2 (Control remains with the entrepreneur)

(i) If in the second stage the entrepreneur is in control, her skill level is greater than one, $\gamma_y > 1$.

(ii) In case of a high profit realization, the entrepreneur obtains a net utility of $u(S^H \gamma_y \Pi^H) - v(e^*)$, while the venture capitalist obtains a monetary payoff of $(1 - S^H)(\gamma_y \Pi^H)$. In case of a low profit realization, the entrepreneur obtains $u(S^L \gamma_y \Pi^L) - v(e^*)$, and the venture capitalist $(1 - S^L)(\gamma_y \Pi^L)$, respectively.

Discussion

The model of Chan, Siegel, and Thakor (1990) investigates the allocation of productive control in venture capital contracting. They explicitly show that a minimum performance is required for the entrepreneur to retain control. Ex post, the allocation of control to either entrepreneur or venture capitalist is socially efficient due to the Pareto-criterion. The venture capitalist's optimal financial claim consists of a risky position in the firm. In case the risk-averse entrepreneur exhibits low management skills, she obtains a fixed remuneration and, thus, gets completely insured against cash-flow randomness. If, by contrast, her skill level is high, the entrepreneur's remuneration is increasing in the project's terminal cash flow.

These findings correspond well to various aspects of real world venture capital contracts. Chan, Siegel, and Thakor (1990), however, do not specify how such abstract variables as management skills or effort level are measured in real life. Second, given the large number of portfolio projects and the restrictive time-budget of venture capitalists, it is rather implausible that a venture capitalist will completely replace the founder and run the project by herself.

4.2.1.2 The search for a professional management

A model very closely related to Chan, Siegel, and Thakor is the analysis provided by Hellmann (1998). Hellmann investigates the venture capitalist's right to appoint a professional outside manager in case the entrepreneur's ability turns out to be below a critical value. Here, the venture capital company does not take control over the project by itself, but rather engages in the search of a superior outside management. This search is associated with increasing search costs and has a success probability smaller than one.

The venture capitalist and the entrepreneur are both risk-neutral. The project outcome can be either a success, yielding a final payoff of one, or a failure, providing a final payoff of zero. The entrepreneur's success probability depends on both the initial skill level and the effort level chosen, $\theta_y(\gamma_y, e_y)$. The success probability of a professional management is higher than the entrepreneur's, $\theta_{PM} > \theta_y$. The entrepreneur derives a private benefit from controlling the project, which is not transferable to other parties. This private, nonpecuniary benefit consists of a fixed component, which is reduced by the entrepreneur's disutility of effort. Thus, the entrepreneur faces a trade-off between the personal benefit and the success probability of the project. Therefore, she may not always act solely in the interest of the venture project. To solve this problem, the entrepreneur must be given an equity stake in the project. However, the equity stake is typically not large enough to provide first-best incentives.

In Hellman's model, the bargaining power is on the side of the venture capitalist. At the end of the game, the venture capitalist pays the entrepreneur a fixed amount, which depends on who is in control of the project. Note, however, that the allocation of control rights is separated from the financial structure, since for any given financial structure it is always possible to allocate control rights independently. Hellmann emphasizes that in venture capital contracting entrepreneurs are, very often, protected too little from being dismissed. The optimal contract in his model induces an excessive rate of replacement. The reason is that entrepreneurs voluntarily relinquish control rights to obtain more favorable financial terms in the contract. Hellmann interprets this result as a self-selection decision of entrepreneurs: Only those willing to cede control rights select venture capitalists for funding the project, while the others seek financing from part of private, more passive investors.

In Hellmann's model, however, no signal is obtained after the first stage which indicates the entrepreneur's management ability. Instead, the venture capital company anticipates potential net benefits from replacing the entrepreneur, which depend on the different project success probabilities, the entrepreneur's private

4.2 Venture capital financing – the individual firm's perspective

benefit, and the slope of the search costs. Taking these into account, the venture capitalist decides whether to engage into the search of a professional management or not. Throughout the model, there is symmetric information. In contrast to Chan, Siegel, and Thakor (1990), it seems indeed more plausible that the venture capitalist provides advice and seeks a professional management rather than controlling the project herself. In our opinion, however, the position of the venture capital company in the model is very strong, maybe too strong given the fact that, empirically, venture capital companies fiercely compete for good project ideas and do not have sufficient bargaining power to throw the entrepreneur out of business.

4.2.1.3 Efficient intervention by the venture capitalist

Marx (1998) presents a model of venture capital contracting in which the contract also specifies the distribution of control rights. The venture capitalist has the right to intervene in the project and to take control of the entrepreneur's project. In contrast to the previous models, Marx explicitly investigates different financial instruments in venture capital contracting and their respective sharing rules. She demonstrates that the optimal contract involves a mixture of both debt and equity and dominates pure-equity or pure-debt financing.

There is a wealth-constrained entrepreneur who owns an innovative project idea, and a venture capital company who seeks profitable investment. Both the entrepreneur and the venture capital company are risk-neutral. The time horizon is one period, which consists of three stages: In the first stage, the contract is signed and the venture capitalist provides the initial investment to start the project. Then, the state of nature is revealed, and the venture capital company decides whether to intervene in the project. Finally, the payoff is realized and distributed to the venture capitalist and the entrepreneur according to the sharing rules.

The random variable z (state of nature), which is continuously distributed in the interval $z \in [\underline{z}; \overline{z}]$ and has a commonly known density and distribution function, positively affects the project's return. The project's return itself is continuously distributed between zero and infinity, $\Pi \in [0, \infty[$. The realized return depends on both the state of nature and the allocation of control rights. Higher z-values imply better states of nature, leading to a first-order stochastic dominance of the return distribution. No effort from part of the entrepreneur or the venture capitalist is required. The project fails if the realized return is insufficient to cover the initial investment. The entrepreneur makes a take-it-or-leave-it offer to the venture capital company, i.e. the bargaining power is on the side of the young firm. The venture capital company signs the financial contract as long as the expected return is nonnegative. The entrepreneur derives a private, nonpecuniary benefit from managing the project. At the beginning of the project, the entrepreneur is in control. However, if a bad state of nature is revealed, the venture capitalist can, at a cost, intervene and take over control. Intervention by the venture capitalist will alter the probability mass of low project outcomes, i.e. it will reduce the failure probability to zero in this range. On the other hand, it has no effect on the probability distribution of high returns, i.e. the expected returns of successful projects are identical when the venture capital company or the entrepreneur is in control. The intuition behind this is that the venture capital company may be able to salvage a struggling project (by providing marketing assistance or good administration); however, it cannot influence the probability of high returns because these are supposed to depend on the quality of the entrepreneur's initial idea. Intervention by the venture capitalist is a rather complex action such that contracts cannot be made contingent upon.

The contract has to specify the sharing rule for the final return. There are three types of financial instruments that can be used to finance the initial investment: debt, equity, and mixed debt-equity. In case of debt financing, the venture capital company obtains a fixed repayment in case of success, and the total, yet insufficient return in case of failure. Under pure-equity financing, the venture capital company obtains a fixed share *(1-S)* of the final return, independent of successful or unsuccessful project realization. Under mixed debt-equity financing, the venture capital company obtains the whole return in case of low payoff realization. In case of high payoff realization, it obtains a fixed payment plus a proportional share for returns in excess of the fixed repayment. Convertible securities typically incorporate the features of this mixed debt-equity sharing rule.

For each financing alternative and sharing rule, Marx (1998) derives the conditions under which the venture capitalist's participation constraint is fulfilled. Then, she calculates the total benefit of the project with and without intervention of the venture capital company. Intervention is efficient if and only if the total benefit is strictly larger with intervention than without. Marx derives that if debt financing is used, the venture capitalist will always intervene when it is efficient and will sometimes intervene when it is not efficient. Under equity financing, the venture capitalist takes control of the company in some states of nature, but not in all where it would be efficient. The reason behind this is the following: Under equity financing, the venture capitalist receives only a fraction of the amount invested from every increase in the project's return and, thus, has no incentive to intervene in states in which it would be efficient. Under debt financing, the venture capitalist receives one dollar for every dollar increase in the project's return if the return is below the initial investment. Thus, when the project's return is low, the marginal benefit from taking control of the venture is greater under debt than under equity financing. Under mixed debt-equity financing (and under the additional assumption that the cost of intervention is sufficiently large) it is shown that the venture capital company takes control only when it is efficient.

Thus, the mixed debt-equity financing and the associated sharing rule maximize the entrepreneur's ex-ante expected utility from the project subject to the venture capitalist's participation constraint. Since this mixed debt-equity contract achieves the first-best solution, there is no scope for contract renegotiation, neither from part of the entrepreneur nor from part of the venture capitalist. Note, however, that the result on the optimality of mixed debt-equity contracts requires that the same investor holds debt and equity and that debt has priority. The inseparability of the fixed component and the proportional component of the mixed debt-equity sharing rule is crucial for the optimality of Marx' result. Once these two features are separated, the holders of the separate financial claims have different (and thus partly inefficient) incentives to intervene in the project.

4.2.1.4 Discussion

The three models presented above all take into account that venture capital investors have the right to intervene in the management decisions of the start-up project. Thus, venture capitalists do not only provide the financial funds for the project's development, but also control or monitor the project in order to reduce failure risk.

Moreover, the approach of Marx (1998) shows that if the contract has to specify both the optimal sharing rule as well as an efficient allocation of control rights, the optimal financial instrument must have two components, i.e. a fixed component and a proportional one. Pure-debt or pure-equity financing alone cannot obtain the first-best solution.

However, given the large number of portfolio projects, the venture capital companies typically do not have the time to control or manage every innovation project. The second group of theoretical papers on venture capital contracting therefore focuses only on the distribution of ownership rights and the associated mechanisms (choice of financial instrument, staged capital infusion) to obtain efficient and incentive compatible investment results.

4.2.2 Allocation of ownership rights

The second group of papers analyzes venture capital financing in the light of the principal–agent theory and focuses on the distribution of ownership rights. The venture capital company as the principal provides equity financing for the R&D investment, while the entrepreneur as the agent has to truthfully allocate these funds. The principal typically cannot observe the investment decision of the agent and, due to this informational asymmetry, a problem of moral hazard arises. Most papers on ownership rights and venture capital contracting investigate a static moral hazard problem (see e.g. Berglöf 1994; Cornelli and Yosha 1997; Trester 1998; Admati and Pfleiderer 1994; and for double-sided moral hazard, Repullo and Suarez 1998; Casamatta 1999, and Schmidt 2000). The first authors who have analyzed dynamical moral hazard in venture capital contracting are Bergemann and Hege (1998). We first briefly describe the static treatments of the principal agent relationship, before we present the dynamical analysis of the moral hazard problem between the entrepreneur and the venture capitalist.

4.2.2.1 Moral hazard and financial instruments

Berglöf (1994) investigates the contractual relationship between a risk-neutral entrepreneur and a venture capitalist and discusses the choice of the financial instruments. The venture project has three potential exit channels: project continuation by the entrepreneur (c), trade sale (s), or liquidation (l). The model's main result is that convertible debt always dominates pure-debt or pure-equity financing or any arrangement combining debt and equity.

The basic model consists of a wealth-constrained entrepreneur who seeks outside capital from a venture capital company. The bargaining power is on the side of the young firm, since there are many investors in the venture capital market but few entrepreneurs with ideas worth investing in. The project is economically viable for one period, which consists of two stages. In the first stage, the contract is signed and the funds are invested. After this, both contracting parties receive a signal about the project's profitability. The state of nature can be either good (which induces high profits) or bad (low profits). The project's return at the end of the period depends on both the state of nature and the respective action taken in the second stage: (c) continuation of the project under the entrepreneur, (s) trade sale to another firm, or (l) liquidation of the firm's assets. The associated profits are observable by both parties, but become verifiable by courts only in case the firm is liquidated. The entrepreneur derives a private benefit B, e.g. reputation, from managing the project, but only when the project is successful. Moreover, Berglöf assumes that in good states of nature, joint profits and utility are highest if the entrepreneur continues to manage the project. However, pure monetary payoffs in the good state are higher if the venture project is sold:

$$\Pi_H^c + B > \Pi_H^s > \Pi_H^c > \Pi_H^l.$$

Here, Π_H indicates the project's profits in good states of nature, while Π_L indicates profits in bad states of nature. In bad states of nature, profits are highest if the firm is liquidated, lower if the entrepreneur continues to mismanage the project, and lowest if the project is sold, because the potential buyer will probably engage in asset stripping:

$$\Pi_L^l > \Pi_L^c > \Pi_L^s.$$

In a next step, Berglöf examines various forms of venture capital contracts: non-voting equity (E), standard debt (D), a combination of standard debt and non-voting equity (DE) or convertible debt (CD). The financing of the initial investment via non-voting equity means that the venture capital company supplies funds in terms of equity and obtains a share of profits in good and bad states. Standard debt financing stipulates a fixed repayment, however, if in bad states of nature the returns are too low, all remaining profits will be transferred to the venture capitalist. Under mixed debt-equity financing, one part of the initial investment is financed via debt while the rest is financed via non-voting equity. The venture capitalist's payoffs consist of the debt repayment plus a share of the remaining profits in the good state of nature, and, if debt repayments cannot be met, all remaining profits in the bad state of nature.[4] In contrast to this, convertible securities imply that the venture capital company signs an option contract: In good states of nature, debt is converted into non-voting equity, and the venture capital company receives $(1-S^{CD})$ of the project's profits. In bad states of nature, the conversion option is not exercised, and the venture capitalist's financial claim is equivalent to the repayment in the standard debt contract.

If there is no external buyer, both the entrepreneur and the venture capitalist agree to continue the project in good states of nature and to liquidate in bad states. This action plan is a constrained first-best solution, which is independent of the choice of the financial instrument.

If an external buyer appears who acquires the firm, there will be an efficiency improvement in good states of nature, since the monetary payoffs increase. In this case, however, the entrepreneur loses her private benefits from running the project. Thus, the arrival of the buyer may improve the situation for the venture capitalist and may worsen it for the entrepreneur, both depending on the financial instrument in place. Given that the venture capitalist breaks even, Berglöf calculates the entrepreneur's expected payoffs under each financial contract and compares the results. In case of a trade sale, standard debt financing will dominate non-voting equity financing, if the efficiency loss from a trade sale in bad states is larger than the expected efficiency improvement by the buyer in good states. (The buyer has to pay a price that gives the reservation utility to the entrepreneur in good states). A combination of debt and equity financing under which the entrepreneur manages the project in good states and the venture capitalist liquidates it in bad states dominates pure-debt or pure-equity financing. To sum up, a convertible debt contract is the best solution since the expected returns to the entrepreneur are monotonically increasing in the fraction of debt.

Discussion

Berglöf's (1994) paper studies different financial instruments in venture capital contracting. He shows that convertible securities, which are widely used in venture capital agreements, dominate the standard instruments of pure-debt or pure-equity financing. Thus, his paper nicely combines theoretical and practical aspects. However, the results are mainly driven by the assumptions made about

---

[4] Here, Berglöf (1994, 255) seems to be wrong: The venture capitalist's payoff under mixed debt-equity financing in good states of nature should be $\theta\,[½(1-S^{DE})(\Pi-½D) + ½D]$ instead of $\theta\,[½(1-S^{DE})\Pi + ½D]$, i.e. the profits should be net of debt repayment.

the ranking of the profits, the signal about the project's profitability, and the entrepreneur's benefits. The assumption that an external buyer always appears is slightly unsatisfactory. Moreover, there is always symmetric information about the state of nature between the entrepreneur, the venture capitalist and the external buyer. Thus, there is no room for moral hazard from part of the entrepreneur, although Berglöf (1994, 262) mentions that "the moral hazard problem can be mitigated by convertible debt". The model would be indeed more convincing if the information distribution was modified to the state of nature being revealed only after the buyer appears and the decision to sell the firm was made. In case information is asymmetric, the investigation of security design will become more complex and some extra incentive constraints have to be included into the model. The paper represents a useful milestone to the question of which financial instruments should be employed in venture capital transactions, but future research should dig deeper into how the project's return is derived.

Hansen (1991) likewise develops a model in which the venture project has different outcomes or exit channels. Here, the project receives two tranches of investment to be completed. After the first period, the entrepreneur – but not the venture capitalist – obtains a signal about the project's profitability. If a clear negative signal is obtained, the entrepreneur will stop the project immediately. If a medium or good signal is received, the entrepreneur will invest the second tranche of capital. Moreover, the entrepreneur has to spend unobservable effort in the second period. Higher levels of effort give rise to "better" final return distributions in the sense that expected returns increase. (The project's profits are, in contrast to the model of Berglöf, continuously distributed). The state of nature, i.e. the quality of the project, is revealed only after the second investment is made and effort is spent. In case the project is good, final profits are distributed between the entrepreneur and the venture capital company according to the sharing rule specified in the contract. If, however, the project turns out to be bad, an additional financing round is required before this high-cost type project is terminated. The risk-neutral entrepreneur maximizes her expected utility from the monetary payoff and the disutility of the effort, subject to the participation constraint of the venture capitalist and the entrepreneur's own incentive constraints. The solution to this maximization problem is a financial contract that combines both debt and equity: The debt component is needed to induce the entrepreneur to provide sufficient effort. The equity component is needed to ensure that the entrepreneur stops a bad project early. The financial instrument that provides the best solution is a convertible security.

The model is clearly written and well structured. Hansen investigates the financial relationship between the venture capitalist and the entrepreneur when contracts can be made contingent on different information constellations. Finally, he derives the equilibrium contract under the very weak assumption that project types are non-verifiable by courts. An nice feature of the model is that bad projects can be sorted out at an early stage.

4.2 Venture capital financing – the individual firm's perspective 129

Similarly in Trester (1998), the quality of the venture project can be of high, medium, and low quality. The model builds on the fact that asymmetric information arises during the project's development, i.e. after the first investment stage the entrepreneur obtains a signal about the project's profitability. This may induce the entrepreneur to moral hazard: If the venture capitalist learns the project's quality only with a time-lag of one period, the entrepreneur can expropriate the project's cash flow for personal consumption. Trester shows that a convertible security dominates pure debt or pure equity financing in mitigating the entrepreneur's moral hazard problem. This means that the venture capitalist has a senior claim on a fraction of the project's return. However, the asymmetric information between the entrepreneur and the venture capitalist is very strong. Moreover, the result of convertible securities being the best financing alternative in this model is based on the assumption that costly auditing to verify the actual cash flow is unavailable in debt contracting. This implies that any optimality of debt is precluded by assumption (see also Thakor's (1998, 701) comment on Trester).

Finally, Cornelli and Yosha (1997) investigate a slightly different moral hazard problem in venture capital contracting: In their approach, the entrepreneur can manipulate the signal about the project's quality (hidden action) in order to obtain refinancing from the venture capital company. The innovation project needs two investment periods. Capital is infused in stages in order to eliminate bad quality projects. The project's returns are continuously distributed between zero and infinity. The venture capital company has the right to stop the project after the first period. The entrepreneur may engage in short-term instead of long-term effort to pretend that the project is of better quality ("window-dressing"). Here, convertible securities also provide the best incentive-compatible financing instrument, which prevent the entrepreneur from exhibiting too favorable intermediary results.

4.2.2.2 Double moral hazard

The works of Repullo and Suarez (1998), Casamatta (1999) and Schmidt (2000) emphasize that venture capital contracting is more than just providing funds for risky innovation projects: Venture capitalists also specialize in giving advice and managerial assistance. They help to select strategies, recruit key-personnel or provide strategic business connections. Thus, both the entrepreneur and the venture capitalist must engage in supporting the project. This particular relationship will be modeled in a double moral hazard setting where both agents must exert an unobservable effort in order to increase the project's profitability. We present the model of Casamatta because it gives us the main idea about the complementarity between the financing and advising roles of venture capitalists. Repullo and Suarez' (1998) approach is more complex: They allow for two investment periods and three types of agents who engage in the start-up project: the entrepreneur, an outside financier and an advisor. Schmidt (2000) also focuses on the incentive effects of convertible securities in solving the double moral hazard problem; his work is very similar to Casamatta (1999).

Assumptions

A1: Players: An entrepreneur owns an innovative product idea and is endowed with some initial wealth A that is invested into the innovation project. The venture capital industry is characterized by perfect competition. The representative venture capitalist has sufficient funds to finance the investment and can provide advisory assistance. Both sides are protected by limited liability.

A2: The time horizon is one period: After the initial funds are invested, both the entrepreneur and the venture capitalist simultaneously spend their effort in order to complete the project. Then the returns are realized and distributed according to the sharing rule specified in the contract.

A3: Information structure: The project can either be a success or a failure, with probabilities θ and $1-\theta$, respectively. The two outcomes Π_H and Π_L are perfectly verifiable. The technical effort spent by the entrepreneur and the advisory effort of the venture capitalist, however, are both unobservable, such that there is mutual asymmetric information.

A4: The entrepreneur and the venture capital company are both risk-neutral.

A5: Production and innovation technology: The success probability θ is linearly increasing in both effort levels: $\theta = \min[e_y + e_{VC}; 1]$, i.e. we assume the two efforts to be additive. Effort is costly, though. The entrepreneur's disutility of effort is given by $v(e_y) = g_y e_y^2 / 2$, while the venture capitalist's disutility of effort is given by $v(e_{VC}) = g_{VC} e_{VC}^2 / 2$. We assume that the effort is less costly for the entrepreneur than for the venture capitalist, $g_y < g_{VC}$. Moreover, for technical reasons, we suppose that $g_{VC} < 2 g_y$ holds.

A6: Due to the reduced form of the project's return, market demand is not explicitly given.

A7: Financial contract: The project requires an investment I. The amount of external financing needed is not fixed but depends on the initial wealth A of the entrepreneur. Thus, the venture capitalist has to supply $I-A$ of funds. The entrepreneur makes a take-it-or-leave-it offer to the venture capitalist. The venture capitalist agrees to participate if she can expect to recoup the monetary and effort investment. There are different financial instruments available: common stock, preferred stock, convertible bonds, and straight debt.

A8: No control rights are distributed.

4.2 Venture capital financing – the individual firm's perspective

First-best solution

The social value of the project Y is given by the expected high and low profits, less the effort costs for the entrepreneur and the venture capitalist, and less the monetary investment:

$$Y(e_y, e_{VC}) = (e_y + e_{VC})\Pi_H + (1 - e_y - e_{VC})\Pi_L - g_y e_y^2/2 - g_{VC} e_{VC}^2/2 - I. \quad (4.17)$$

Under full information, i.e. when both efforts are observable, the optimal effort levels are given by the first-order condition of maximizing Y:

$$e_y^{FB} = \frac{1}{g_y}(\Pi_H - \Pi_L),$$

$$e_{VC}^{FB} = \frac{1}{g_{VC}}(\Pi_H - \Pi_L), \quad (4.18)$$

where FB indicates first-best. In order to make things interesting, i.e. to have a success probability θ smaller than one, we need the additional assumption that the sum of weighted efforts of each agent multiplied with the difference in profit levels must be lower than one, $\left(\frac{1}{g_y} + \frac{1}{g_{VC}}\right)(\Pi_H - \Pi_L) < 1$. Inserting the first-best effort levels into (4.17), the first-best value of the project is given by:

$$Y^{FB} = \tfrac{1}{2}\left(\tfrac{1}{g_y} + \tfrac{1}{g_{VC}}\right)(\Pi_H - \Pi_L)^2 + \Pi_L - I. \quad (4.19)$$

The first-best equilibrium requires both agents to exert efforts. This implies that if there is no moral hazard problem, it is optimal for the entrepreneur to ask for the services of an advisor. However, the entrepreneur's effort level exceeds the venture capitalist's one, $e_y^{FB} > e_{VC}^{FB}$, since we have assumed that the venture capitalist's effort is more costly.

Optimal contract with moral hazard

When efforts are unobservable, each participant in the contract chooses an effort level which maximizes her expected utility. The financial contract maximizes the expected utility of the entrepreneur, subject to the incentive compatibility conditions of both agents and the participation constraint of the venture capitalist. The financial contract allocates the shares $\{S^H, S^L\}$ to the entrepreneur when profits are $\{\Pi_H, \Pi_L\}$, while the venture capitalist obtains the shares $\{1\text{-}S^H; 1\text{-}S^L\}$.

The maximization program of the entrepreneur is as follows:

$$\max_{e_y, S^H, S^L} V = (e_y + e_{VC})S^H \Pi_H + (1 - e_y - e_{VC})S^L \Pi_L - g_y \frac{e_y^2}{2} - A. \quad (4.20)$$

Thus, the entrepreneur maximizes her share of expected profits, incurs effort costs and is willing to spend her entire wealth A on the project.

The venture capital company's participation constraint has to satisfy:

$$W = (e_y + e_{VC})(1-S^H)\Pi_H + (1-e_y - e_{VC})(1-S^L)\Pi_L - g_{VC}\frac{e_{VC}^2}{2} \geq I - A. \quad (4.21)$$

Note that the amount of external financing required from the venture capitalist depends on the difference between the investment needs and the size of the entrepreneur's initial wealth. Lastly, due to the limited liability of both the entrepreneur and the venture capitalist, profit shares are restricted to:

$$\{S^H; S^L\} \in [0,1]. \quad (4.22)$$

The optimal effort levels under asymmetric information are derived from the first order conditions of the incentive compatibility constraints. Given that $\frac{1}{g_y}\Pi^H / g < 1$, the optimal effort levels are specified as:

$$\frac{\partial V}{\partial e_y} \stackrel{!}{=} 0 \Rightarrow e_y^* = \frac{1}{g_y}(S^H\Pi_H - S^L\Pi_L),$$

$$\frac{\partial W}{\partial e_{VC}} \stackrel{!}{=} 0 \Rightarrow e_{VC}^* = \frac{1}{g_{VC}}((1-S^H)\Pi_H - (1-S^L)\Pi_L). \quad (4.23)$$

Each agent's effort level will be higher the greater the difference between high and low profit levels. Moreover, the entrepreneur's effort is increasing in S^H and decreasing in S^L. Likewise, the venture capitalist's effort is increasing in $(1-S^H)$ and decreasing in $(1-S^L)$. The special nature of the incentive problem is that the two agents must share a fixed final outcome. This implies that the more powerful the incentives given to one agent, the less the other agent will be induced to increase effort. The optimal contract has to take these countervailing effects into account.

Proposition 4.3 (No external financing need)

When there is no need for external financing, i.e. I-A=0, the venture capitalist exerts no effort ($e_{VC}=0$) and the entrepreneur exerts the first-best level of effort (e_y^{FB}).

Proposition 4.3 builds on the fact that the venture capitalist's participation constraint is always binding: If the venture capitalist does not invest any money in the project, she will obtain no share of the final outcome, which leads to no effort e_{VC} being exerted. The entrepreneur then chooses the optimal effort level, but the

4.2 Venture capital financing – the individual firm's perspective

value of the project is strictly lower than first-best Y^{FB}, because joint efforts are necessary to implement the first-best solution. If the entrepreneur wants to make the venture capitalist participate as an advisor only (and not as a financial investor), the share of profits that has to be transferred is too large compared to the increase in expected value derived from the expert's advice.

Proposition 4.4 (Optimal contract)

When the entrepreneur needs external financing, i.e. I-A>0, there exists a unique optimal contract $\{S^H, S^L\}$ defined by

- $S^H = \frac{1}{\Pi_L}[\Pi_L + \frac{(g_{iv} - g_y)(\Pi_H - \Pi_L) + g_y g_{iv} \sqrt{\Omega_1}}{2 g_{iv} - g_y}]$
  $S^L = 1$ $\quad$ *if $(I - A) \in]0, \Psi]$,*

- $S^H = \frac{1}{\Pi_H}[\Psi + \Pi_L - (I - A) + \frac{g_{iv}(\Pi_H - \Pi_L)}{g_{iv} + g_y}]$
  $S^L = \frac{1}{\Pi_H}[\Psi + \Pi_L - (I - A)]$ $\quad$ *if $(I - A) \in]\Psi, \Psi + \Pi_L]$,*

- $S^H = \frac{1}{\Pi_H}[\frac{(g_{iv} - g_y)(\Pi_H - \Pi_L) + g_y g_{iv} \sqrt{\Omega_2}}{2 g_{iv} - g_y}]$
  $S^L = 0$ $\quad$ *if $(I - A) \in]\Psi + \Pi_L; \bar{I}]$;*

where $\Psi = \frac{(g_y^2 + 2 g_{iv}^2)(\Pi_H - \Pi_L)^2}{2 g_{iv} (g_{iv} + g_y)^2}$ is a constant that separates the investment regions.[5]

In Cassamatta, the optimal contract depends on the size of the external investment required:

When the external financing need *(I-A)* is small, it is optimal not to leave any share of the low profits to the venture capitalist in order to induce her to exert effort. The participation constraint is satisfied with the venture capitalist's expected share of high profits.

When the demand of external financing is large *(I-A > $\Psi+\Pi^L$)*, the participation constraint is more difficult to fulfill, and the venture capital company must obtain a large share of the final outcome in both states of nature. At the same time, the entrepreneur must exert effort. For incentive reasons, the optimal contract specifies that the entrepreneur obtains S^H in god states of nature, while in bad states she obtains $S^L = 0$.

For a medium investment level, both agents' effort levels are constant. The value of the project remains constant too, however, under double moral hazard, the optimal project value Y^* is always lower than the first best Y^{FB} (see Figure 4.2):

---

[5] See Casamatta (1999, Appendix, 26-28), Ω_1 and Ω_2 are functions of (I-A).

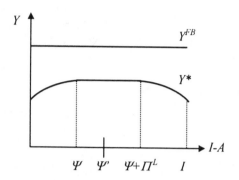

Figure 4.3: Project value under venture capital financing and double moral hazard
(Adapted from Casamatta 1999, 12)

Figure 4.3 indicates that Y^* is increasing with $I-A$ until Ψ, because in this interval it is less costly to share effort between the two agents than to have effort exerted by the entrepreneur only. Y^* reaches a maximum if the amounts invested by each party are rather balanced and induce sufficient effort levels e_y^* and e_{VC}^*. The entrepreneur is willing to always have a financial partner investing in the project, even though she would be wealthy enough to implement the project alone. If $I-A$ becomes too high, large shares of profits have to be transferred to the venture capitalist, thus, e_{VC}^* becomes too high, e_y^* too low, such that Y^* begins to decrease.

Implementing the optimal contract

What are the financial instruments to implement the optimal contract? The choice of the financial instrument depends on the level of external financing required:

When external financing is rather small, $I - A \in [\Psi, \Psi']$, with $\Psi' = \Psi + \frac{g_y}{g_y + g_{vc}} \Pi_L$, the venture capitalist should be given common stocks and the entrepreneur should obtain preferred stocks. Preferred stocks ensure a minimum rate of return to their owner before common stock's returns are paid. When the project outcome exceeds a certain threshold, common stocks and preferred stocks will generate the same rate of return. When the outcome of the project does not reach this level, it is impossible to remunerate common stocks with the same rate of return as preferred stocks. Thus, in case of low profit realization, the entrepreneur obtains a large share of profits for incentive reasons when the external financing needs are low.

4.2 Venture capital financing – the individual firm's perspective 135

When the amount of external investment is substantial, i.e. $I - A \in [\Psi', I]$, the financing instrument should be designed as a convertible bond: Here, the venture capitalist as external investor obtains a large share of the project outcome in bad states of nature. The entrepreneur is better remunerated in good states of nature, which makes her intensify her effort to increase the probability of high profits. When profit realization is high, bonds are converted into common stocks, and the venture capital company obtains a share of these high profits. It is worth noting that preferred stocks are equivalent to convertible bonds, as long as preferred stocks ensure a minimum revenue to be given to their owner. Differences between these two types of financial instruments usually concern the right to trigger bankruptcy, which, in case of convertible bonds, is given to the venture capitalist.

The results state that when the venture capitalist's participation is low, she should obtain common stocks, while she obtains convertible bonds or preferred stocks when the investment is high. These results are consistent with the empirical findings of Fenn, Liang, and Prowse (1998).

Discussion

The previous analysis stresses some important features of the financing of start-up firms. In particular, when the entrepreneur lacks experience and know-how concerning business administration, a financial partnership with an investor-advisor is desirable, so that the two agents exert their competencies effectively. Involving an "active" investor into the project enhances its value, which can explain why venture-capital backed firms exhibit greater revenue growth than other similar firms. The double moral hazard problem is partly solved by the two agents' financial contributions. By focusing on the necessary joint efforts of the entrepreneur and the venture capitalist to implement the project efficiently, the model offers a possible explanation for the common use of convertible securities in venture capital contracting. The double moral hazard model differs from the allocation of control rights models (section 4.2.1) in that the venture capitalist here supports the project over the entire life span, instead of making just one single control decision. The model is very nice, because with simple assumptions it obtains very clear results.

4.2.2.3 Contracting with a venture capitalist and an outside investor

As we can infer from the financial growth cycle paradigm, the participation of a venture capital company can be interpreted as a quality signal to outside investors: If a venture capitalist has selected to invest in a certain project, then this innovation project's prospects are presumably good, and banks might be willing to invest into this project at a later stage. The investment decision of the venture capitalist reduces the initial risk of the pool of projects, because the venture capital company scrutinizes every project via a tough evaluation process. This due diligence procedure helps to mitigate the acute information problem between external investors and the entrepreneur. Therefore, in later stages of an innovation project, other financial intermediaries become interested in taking part in the project, too. Admati and Pfleiderer (1994) investigate how a financial contract is designed if a third party, i.e. an outside investor like a bank, invests into the innovation project after the venture capitalist has already decided to take a stake in the young firm.

Assumptions:

A1: Players: An entrepreneur has an innovative project idea, but no initial wealth. The entrepreneur seeks external financing and is subject to limited liability. A venture capital company and an outside investor decide whether to invest in the innovation project. All capital markets are characterized by perfect competition.

A2: The time horizon is two periods. In each period, the project requires a monetary investment I_1, I_2 for development. The project will be abandoned if a negative signal is obtained after the first period. The venture capital company provides the first-period investment, whereas the outside investor and the venture capital company jointly finance the second-period investment. Final payoffs are realized only at the end of the second period.

A3: Information structure: The outcome of the project can either be a success or a failure. When the project succeeds, the final payoff equals Π, whereas when it fails, the payoff is zero. The probability of success θ_z is initially unknown and depends on the state of nature $z \in [\underline{z}; \overline{z}]$. Better states of nature, i.e. higher z-values, imply a higher success probability θ_z. After the first period the state of nature is revealed to both the entrepreneur and the venture capitalist, but not to the outside investor. In the second period, thus, an informational asymmetry exists between the outside investor on the one side and the entrepreneur/venture capitalist on the other side. The intuition behind this is that the venture capitalist works closely with the firm, monitors it frequently and is generally well informed about the firm's prospects. The outside investor, by contrast, solely provides money for the later stage investment and does not observe the insider's information. Although the state of nature is observable by both the entrepreneur and the venture capitalist, it cannot be verified by courts. The initial financial contract, therefore, cannot be based upon the realization of z and cannot prescribe an automatic stop of the project after a negative signal is obtained.

A4: Risk preferences: All players are risk-neutral.

A5: The production technology is not fully specified. If the project is abandoned after the first period, a random liquidation value $L_z > 0$ can be recovered which is determined by the state of nature. If the project is continued, the return at the end of the second period is continuously distributed and depends on both the second-period investment and the success probability, $(\Pi | I_2, \theta_z)$.

A6: Due to the reduced form of the project's returns, market demand is not explicitly given.

A7: Financial contract: In the first period, the entrepreneur and the venture capital company enter into an initial financing agreement. The entrepreneur sells shares sufficient to raise capital I_1 from the venture capitalist. The initial financial contract specifies *(i)* the sharing rule according to which the liquidation value is distributed if the project is abandoned after the first period, and *(ii)* the set of financing arrangements that are permissible for second-period financing. If the project is continued, an amount I_2 is raised, of which the venture capitalist will provide ξI_2 and the outside investor will provide *(1-ξ) I_2*. The entrepreneur wishes to set up the initial financial contract such that *(i)* the venture capital company makes an optimal continuation decision and *(ii)* has incentives to choose a second-period investment level that maximizes the ex-ante value of the project.

A8: No control rights are officially distributed, but the venture capitalist implicitly controls the decision whether to continue the project or to abandon it after the first period.

Before we derive the optimal financial contract, we analyze the conditions under which the project is continued after both entrepreneur and venture capitalist have learned the state of nature.

Project abandonment or continuation decision

Let $z_{low} \in [\underline{z}, \hat{z}]$ be the set of states for which it is optimal to abandon the project. That is, $z \in z_{low}$ if and only if

$$\sup_{I_2} [E(\Pi | I_2, z) - I_2] < L_z. \tag{4.24}$$

In these states of nature, the liquidation value is greater than the net expected profits (after the initial investment I_1 is sunk), and it is first-best not to continue the project.

The initial financing agreement induces optimal continuation if and only if for any possible state of nature, i.e. for every L_z and for any family of distributions $F(.|I_2,z)$, the following holds:

(i) If $z \in z_{low}$, then

$$\sup_{I_2, \xi} [(1-S)E(\Pi | I_2, z) - \xi I_2] \leq (1-s)L_z. \tag{4.25}$$

Condition *(i)* says that if $z \in z_{low}$, i.e. if the optimal strategy is to abandon the project, then the venture capital company will at least weakly prefer that no continuation takes place: The venture capitalist's share of expected profits, net of the fraction of second-period investment that she finances, is smaller than her respective share of the liquidation value.

(ii) If $z = z_{high} \in [\hat{z}, \bar{z}]$, then the project is continued and there exists a second-period financial contract such that

$$I_2 \in \arg\max_{I_2} [E(\Pi|I_2, z) - I_2] ; \qquad (4.26)$$

$$[(1-S)E(\Pi|I_2, z) - \xi I_2] \geq (1-s)L_z ; \qquad (4.27)$$

$$(1-S)E(\Pi|I_2, z) - \xi I_2 \geq (1-S')E(\Pi|I'_2, z) - \xi' I'_2$$
$$\forall \{\xi', I'_2, (1-S')\}. \qquad (4.28)$$

Condition *(ii)* states that if $z \in z_{high}$, i.e. if continuation is optimal, then there is a permitted financing arrangement for the second period $\{\xi, I_2, (1-S)\}$ that involves the optimal level of investment I_2 (4.26). Moreover, this financial contract makes the venture capitalist choose to continue rather than to abandon the project (4.27), and it leads to the highest payoffs to the venture capitalist among all possible financing arrangements (4.28).

Optimal financial contract between entrepreneur and venture capitalist

However, there may exist various financing arrangements that induce an optimal continuation decision by the venture capitalist. Admati and Pfleiderer identify a particular type of contract which involves optimal project continuation, and which is robust to changes such as adding another possible state of the world. This robust optimal contract is a fixed-fraction contract with the following features: The venture capitalist finances the first-period investment and obtains a fraction $(1-S)$ of the project's total payoffs no matter what the continuation strategy is: In case of abandonment, the venture capitalist obtains the share $(1-s) = (1-S)$ of the liquidation value. If the project is continued and a second-period investment I_2 is needed, then the venture capitalist invests $\xi I_2 = (1-S)I_2$ and contacts an outside investor, who finances the rest, $(1-\xi) I_2$. Thus, the fraction of the second-period investment supplied by the original venture capitalist is equal to the fraction of the total payoffs the venture capitalist would obtain if no continuation took place. Moreover, this is the same fraction of total payoffs that the venture capitalist obtains for any positive second-period investment level I_2. The following proposition summarizes the necessary and sufficient conditions for the set of initial financing arrangements which induce optimal project continuation.

Proposition 4.5 (Fixed-fraction contract)

The initial financial contract between the entrepreneur and the venture capitalist induces optimal project continuation if and only if there exists a fixed share $(1-S) \in (0,1]$ such that

(a) the venture capitalist's fraction of the liquidation value is $(1-s)L_z = (1-S)L_z$;

(b) for all positive second-period investment levels $I_2 > 0$, the venture capital company finances the fraction $\xi = (1-S)$ and obtains the share $(1-S)$ of expected payoffs, $\theta_z \Pi$;

(c) for all states of nature z and all financial arrangements, any other contract than the fixed-fraction contract is redundant, because the fixed-fraction contract gives the smallest payoff to the venture capitalist $E[(1-S)(\Pi | I_2, z)] - \xi I_2 \leq (1-S)[E(\Pi | I_2, z) - I_2]$, which in turn generates the highest surplus for the entrepreneur.

To obtain some intuition for the result, we restrict the possible contracts to be linear in the total payoff. If, for example, the fraction of the second-period investment financed by the venture capitalist were higher than the share of final payoffs, the venture capitalist would sometimes abandon a project that should be continued. On the other hand, if the venture capitalist obtained a higher share of payoff than the fraction of the second-period investment, then there would exist projects that the venture capitalist would choose to continue, although it would be better to abandon them.

When the outside investor is involved in the second-period financing, a fixed-fraction contract helps to resolve the informational asymmetry between the entrepreneur/venture capitalist and the uninformed outside investor: This follows because with a fixed-fraction contract, the venture capitalist's payoff is independent of the pricing of any securities issued to the outside investor. The venture capitalist, being a shareholder from the first period, has incentives to overprice new securities because this increases the project's returns. However, being also a new capital provider for the second-period investment, the venture capitalist likewise has an interest to sell the young firm's shares cheaply and to underprice new securities. Admati and Pfleiderer point out that with a fixed-fraction contract the venture capitalist's incentives to overprice or underprice securities are exactly offset. Consequently, the venture capitalist will reveal the information about the state of nature truthfully. The venture capitalist has no incentive to misrepresent it, because she obtains the same share of payoff independent of the continuation decision and the state of nature. The venture capitalist, therefore, plays a kind of certification role to the outside investor in the issuance of new securities.

Note that the fixed-fraction contract is similar to equity financing, because it gives the venture capitalist a fixed share of the payoff. Unlike common equity, however, the venture capitalist does not own a residual claim, since her shares have priority over all other claims issued by the firm. Therefore, we can interpret this financial instrument as a preferred equity, in which the venture capitalist's claim has priority over all other payoff distributions.

Discussion

Admati and Pfleiderer (1994) discuss a model in which a young, innovative firm issues securities to finance its early-stage and later-stage investments. A venture capitalist buys first-period shares to finance the initial investment. Due to the close relationship with the entrepreneur, the venture capitalist gets access to the same state of knowledge as the entrepreneur about the project's profitability. This information, however, is not revealed to outside investors who are involved in the later-stage investment. The entrepreneur could, therefore, exploit this informational asymmetry and overprice the second-period securities, which would generate higher final payoffs. If, instead, the venture capitalist as an inside investor decides upon the issue price of the later-stage securities, the firm's value will be conveyed truthfully to outside investors, i.e. the securities will be priced correctly, if and only if the venture capitalist holds a fixed fraction of the second-period securities herself and is entitled to the same fixed fraction of the final payoff and the same fixed fraction of the liquidation value. Thus, the venture capitalist plays an important role to bridge the informational gap between inside and outside investors.

Admati and Pfleiderer's (1990) approach is very interesting because it combines early-stage and later-stage financing of the innovation project and analyzes the financial involvement between various types of investors. The model fits very well with reality in that the venture capitalist is the first or lead investor. Venture capitalists specialize in investing in new projects which are characterized by massive uncertainty. Besides frequent monitoring and information gathering, venture capitalists take an active part in management decisions such as whether to stop or to continue the project and how to raise additional capital contributions. The outside investor, by contrast, steps in only after some initial project risk is reduced, i.e. after he observes that the project is continued, which serves as public signal that the project is worth funding.

The contractual arrangement resembles a delegation problem with two agents in the sense of Melumad, Mookherjee, Reichelstein (1995): The entrepreneur, as the principal, delegates the decision whether to continue the project and how to price the second-period securities to the venture capitalist. The venture capitalist, being the first agent, subcontracts with an outside investor, the second agent, who finances the remaining fraction of the second-period investment. This organizational design combined with the fixed-fraction contract helps to bridge the informational gap between inside and outside financiers.

However, the approach of Admati and Pfleiderer (1994) is problematic in two ways: First of all, it is unclear how the outside investor is compensated for his second-period investment. The final return is distributed only between the entrepreneur and the venture capital company. This means that the venture capitalist cannot repay the outside investor out of her profits, because this contradicts the fixed-fraction contract. A rational outside investor, therefore, would never be willing to participate in the second-period investment. Thus, the only solution is that the outside investor continues to hold the young firm's securities beyond the end of the game, and that the firm's shares can be traded after the game is over. This, however, is *not* formalized in the model, i.e. the equilibrium robust financial contract is not subgame-perfect!

Our second problem concerns the continuation decision: Admati and Pfleiderer state that the higher the liquidation value, the less likely is project continuation. This trade-off, however, seems to be at odds with reality: A high liquidation value will, generally, be realized if the firm's tangible assets are high. High tangible wealth, in turn, indicates that the project's prospects are rather good. Therefore, Admati and Pfleiderer should relate the liquidation value positively to the final payoff, instead of creating an artificial trade-off. A second solution is to reinterpret the "liquidation value" as a benefit which will be created if the young firm fails, the project is sold, and another firm of the same industry makes use of the research and development results.

Lastly, a remark concerns the information structure: While it is true that a venture capitalist is generally well informed about the activities of her portfolio companies, we believe that an informational asymmetry still exists between the investor, who is interested mostly in cash flow realization, and the entrepreneur, who is the expert for the technical development.

Therefore, in the next subsection, we turn back to a model that investigates financial contracting under asymmetric information, and analyzes the ensuing moral hazard problem between the entrepreneur and the venture capitalist. The analysis is carried out in a dynamical context.

4.2.2.4 *Dynamic contracting, moral hazard, and learning*

The model of Bergemann and Hege (1998) is characterized by a long-term financial relationship, in which venture capital is provided for developing a multi-stage innovation project. The allocation of funds is subject to dynamic moral hazard from part of the entrepreneur. The project's quality is initially unknown and more information arrives after each development phase. This research and development process is stopped if either the innovation project is successfully realized or if expected future returns drop below the innovation costs of that period.

Assumptions

A1: Players: An entrepreneur has an innovative project idea, but no initial wealth. A venture capital company seeks profitable investment into a portfolio company. The venture capital industry is characterized by perfect competition.

A2: The time horizon consists of $t=1,..T$ periods. In each period, the venture capital company supplies funds to finance the research and development, while the entrepreneur has to truthfully allocate these venture funds into the innovation project. Since the project horizon is long-term, future periods are discounted by the factor δ.

A3: Information structure: There are three uncertainties in the model.
(i) The project quality is initially unknown. Both parties share common priors that the project's quality is good with probability α_1 and bad with probability $1-\alpha_1$. Good quality projects succeed with probability θ_1 and fail with probability $1-\theta_1$. Bad projects fail with certainty. At the end of each period, both parties update their beliefs about the project's quality according to the Bayes-rule.
(ii) Asymmetric information about the investment of funds: The venture capital company is unable to observe whether the entrepreneur truthfully invests the money towards the discovery process or whether she diverts them to her private ends.
(iii) Asymmetric information in the learning process: If the entrepreneur truthfully invests the funds, but the project fails in the present period, both the entrepreneur and the venture capital company share the same a-posteriori beliefs about the project's quality. If, however, the entrepreneur misuses the funds, she knows that the project failed due to her moral hazard behavior. Thus, while the venture capitalist reevaluates the project's quality, the entrepreneur does not update her beliefs, and asymmetric information arises in the learning process. To prevent this informational asymmetry, the entrepreneur gets access to an information rent.

A4: Risk preferences: The entrepreneur and the venture capitalist are both risk-neutral.

A5: Production and innovation technology: The innovation process is associated with research and development costs. These R&D costs are linear in the success probability: $g(\theta_t)=g\theta_t$, with g denoting a positive cost parameter. In case of success, the project yields an exogenous return of Π, which is assumed to exceed the investment costs by many times.

A6: Due to the reduced form of the project's return, market demand is not explicitly given.

A7: Financial contract: The entrepreneur makes a take-it-or-leave-it offer to the venture capital company. Thus, the bargaining power is on the side of the young firm, and the venture capital company signs the financial contract as long as her expected return is nonnegative. The optimal contract is a time-varying share contract. At the same time, it represents an option contract, because the entrepreneur chooses either to take the money and run, or to obtain a share of the project's returns.

A8: No control rights are distributed, it is always the entrepreneur who manages the project.

Information acquisition and learning process

As the experimentation process develops over time, all players learn more about the project's prospects. If success has not yet occurred at period t, both the entrepreneur and the venture capitalist update their beliefs about the project's quality. The evolution of the a-posteriori beliefs is derived according to the Bayes-rule: a_{t+1} denotes the posterior belief that the project is good, conditional on the fact that no success has been realized until period t. The a-posteriori beliefs a_{t+1} are a function of the present period's beliefs α_t and the amount of R&D investment $g\theta_t$, (implicitly, of the success probability θ_t):

$$\alpha_{t+1} = \frac{\alpha_t(1-\theta_t)}{\alpha_t(1-\theta_t)+1-\alpha_t} = \frac{\alpha_0 \prod_{s=0}^{t}(1-\theta_s)}{\alpha_0 \prod_{s=0}^{t}(1-\theta_s)+1-\alpha_0} \qquad (4.29)$$

The left-hand side of equation (4.29) states that the future belief of investing into a good-quality project consists of the probability that a good project fails in the present period divided by the total failure probability of the present period. By solving this recursively, we obtain the term on the right-hand side which represents the development of the a-posteriori beliefs in period $t+1$ as a function of the initial beliefs α_0. The beliefs α_{t+1} decrease over time if no discovery has been made.

Continuation decision and total project value

The social value of the venture project is maximized by choosing the optimal level of R&D investment (i.e. by determining indirectly the success probability θ_t) and by selecting the optimal stopping period, T^*.

The research project is stopped as soon as the current expected profits fall below the investment costs:

$$\alpha_t \theta_t \Pi - g \theta_t \leq 0; \qquad \Rightarrow \qquad \alpha^* = g / \Pi . \qquad (4.30)$$

It follows that the posterior beliefs α^* at which efficient stopping occurs decrease when either the project's return increases or the marginal research costs g decrease.

If indeed the last investment occurs at a_T, then it is optimal to invest the maximal amount $g\theta^{max}$ in the final period. This is due to the linearity of the research cost function. Hence, we obtain for the project value in the terminal period

$$Y(\alpha_T) = \alpha_T \theta \Pi - g \theta^{max} . \qquad (4.31)$$

The total value of the project is, then, derived recursively by the dynamic programming equation:

$$Y(\alpha_t) = \max_\theta \{\alpha_t \theta_t \Pi - g \theta_t + (1 - \alpha_t \theta_t) \delta \cdot Y(\alpha_{t+1})\}, \qquad (4.32)$$

where δ represents the discount factor. Higher research expenditures will raise the success probability in the present period and the associated expected returns $\alpha_t \theta_t \Pi$. At the same time, it becomes less likely that the project is continued in the subsequent period as $(1 - \alpha_t \theta_t)$ decreases and the continuation value $Y(\alpha_{t+1})$ declines.

Proposition 4.6 (Optimal investment policy in the multi-period innovation project)

(i) It is optimal to invest the maximal amount $g\theta_t^{max} = g\theta$ into the research project until T^*.

(ii) Under this investment policy, the social value of the venture project equals:

$$Y(\alpha_t) = \sum_{s=t}^{T^*} \delta^{s-t} [\alpha_t \theta (\Pi - g)(1 - \theta)^{s-t} - (1 - \alpha_t) g \theta]. \qquad (4.33)$$

The social value of the project is given by the sum of discounted net profits when the project is successfully realized in $t = 1, 2, \ldots T^*$, less the research expenditures spent on bad quality projects. By reformulating equation (4.33) in terms of the initial beliefs α_0, we obtain the present value of the innovation project in terms of a geometric series:

$$Y(\alpha_0) = \alpha_0 \theta (\Pi - g) \frac{1 - \delta^{T^*}(1-\theta)^{T^*}}{1 - \delta(1-\theta)} - (1 - \alpha_0) g \theta \frac{1 - \delta^{T^*}}{1 - \delta} . \qquad (4.34)$$

The first term represents the net value of the venture project conditional on the project being good, while the second term represents the discounted value of funds misallocated on bad projects.

Financial contracting

We now turn to the financial arrangement between the entrepreneur and the venture capitalist. The venture capital company supplies funds for the innovation project in exchange for a share of the uncertain return. As stated above, the venture capital company cannot observe whether the entrepreneur truthfully allocates the funds or not. The entrepreneur maximizes her expected share of profits, given that the venture capitalist at least breaks even. The entrepreneur mustn't have any incentive to divert the funds. We begin by analyzing the provision of venture capital under short-term contracts.

Short-term contracts

S_t shall denote the share of profits accruing to the entrepreneur if the project succeeds in period t, i.e. Π is realized. $(1-S_t)$ is the corresponding share of the venture capitalist. The venture capitalist's participation constraint in the current period t equals:

$$(1-S_t)\alpha_t\theta_t\Pi = g\theta_t. \qquad (4.35)$$

At the same time, the remaining share S_t for the entrepreneur has to be high enough to prevent her from diverting the funds. The entrepreneur's incentive compatibility constraint is, therefore, given by:

$$S_t\alpha_t\theta_t\Pi \geq g\theta_t. \qquad (4.36)$$

By combining these two restrictions (4.35) and (4.36) we see that short term financing can be continued as long as the inequality

$$\alpha_t\theta_t\Pi \geq 2g\theta_t \qquad (4.37)$$

is satisfied. We infer that the project is stopped if the critical value of the posterior belief drops below $\alpha_t \leq 2g/\Pi$. By comparing this to the socially efficient stopping point $\alpha^* \leq g/\Pi$, we derive that the funding horizon under short-term contracting T^{ST} is shorter than the efficient horizon T^{FB}. It implies that the venture project is stopped prematurely under short-term contracting.

Thus, short-term financing leads to inefficiencies which are due to the information problem and the conflict of interest between the entrepreneur and the venture capitalist: The competing claims arising from the investment problem of the financier and the agency problem of the entrepreneur impose serious restrictions on the venture project's profitability. As the posterior beliefs decrease over time, it becomes more difficult to cover both parties' compensations from the expected returns in period t. As a consequence, short-term financing will terminate the innovation project too early. Bergemann and Hege presume that a long-term contract, which allows for intertemporal transfers, will generally be more efficient.[6] In the next section, therefore, we investigate the financial relationship if both parties can commit to long-term contracting.

Long-term contracting

Under a long-term financial contract, the venture capital company supplies funds for multiple investment rounds. The main difference to short-term contracting is that the venture capitalist needs not to break even in each period. Instead, it suffices when the venture capital company obtains nonnegative expected profits over the total funding horizon. Technically speaking, the sequence of static participation constraints is replaced by a single intertemporal participation constraint. As a consequence, the share of potential profits $(1-S_t)$ accruing to the venture capitalist in each period can be reduced. On the other hand, the dynamic financial contract offers more opportunities for the entrepreneur to divert the funds. The entrepreneur can, for example, misuse the funds today and bet on a positive project realization tomorrow. This implies that, especially in early periods, the entrepreneur's share S_t must be rather large to prevent her from diverting the funds. We begin by analyzing the moral hazard problem in the final period:

The entrepreneur's expected share of payoffs must be larger than the capital supplied for research expenditures:

$$S_T \alpha_T \theta_T \Pi \geq g \theta_T . \tag{4.38}$$

The minimal share S_T in the ultimate period is then given by $S_T = g/\alpha_T \Pi$. Solving the minimization problem recursively, we obtain the expected value $V_T(\alpha_t)$ that the entrepreneur receives for a given funding policy:

$$V_T(\alpha_t) = S_t \alpha_t \theta_t \Pi + \delta(1 - \alpha_t \theta_t) \cdot V_T(\alpha_{t+1}) . \tag{4.39}$$

It consists of the share of expected profits if the project succeeds in the present period t, and the discounted value if the project succeeds in future periods. The entrepreneur, however, can misuse the funds $g\theta_t$ with the following three consequences: First of all, the entrepreneur enjoys utility from diverting these funds $g\theta_t$ for private consumption. In this case, the success probability drops to zero and the project fails in the present period. Second, if the project fails due to the withholding of funds, the venture capitalist updates her beliefs about the

---

[6] The statement that "... efficient financing requires some form of *intertemporal risk sharing* ..." (Bergemann and Hege, 1998, 714) is, however, wrong since both the entrepreneur and the venture capitalist are risk-neutral.

project's quality. The entrepreneur, on the other hand, knows that the project failed because of her opportunistic behavior and does not update her beliefs about the project's quality. To fend off this informational asymmetry, the entrepreneur is granted an information rent. This rent is higher the more control the entrepreneur exerts over the conditional probability θ_t, i.e. the larger the amount of fund is needed. Third, the discounting of future returns at the rate δ reduces the expected value of the project if the innovation is realized in later periods.

Proposition 4.7 (Share contract)

The minimum incentive-compatible share S_t that the entrepreneur must obtain for truthfully investing the funds is determined by the following three aspects:

- *The incentive constraint of the present period (static moral hazard costs);*

- *The intertemporal agency problem: The entrepreneur has to be compensated for not misusing the funds in a single period and continuing afterwards as instructed until T.*

- *The learning rent: The entrepreneur has to be compensated for not manipulating the development of beliefs.*

The development of the minimum shares S_t over time is determined by an underlying option problem: The entrepreneur has the option to withhold the funds in a single period and to consume the amount $g\theta_t$ right away, or the entrepreneur can expect to obtain the share of profits $S_t\Pi$.

Proposition 4.8 (Evolution of shares over time)

The optimal financing contract is a time-varying share contract. For a discount factor sufficiently close to one, the entrepreneur's share of profits S_t is monotonically decreasing.

The equilibrium contract and the implied funding horizon are determined by integrating the intertemporal incentive constraint and the intertemporal participation constraint. Bergemann and Hege show that a long-term contract allows for an extended funding horizon (compared to short-term contracts). However, a long-term contract still doesn't attain the efficient length of financing, T^{FB}, because of the dynamic agency problem and the asymmetric information.

Solutions to the early stopping problem

In order to circumvent the early stopping problem, Bergemann and Hege point out the following remedies: The model can be extended by a positive liquidation value; or the venture capital company engages in monitoring activities.

a) Liquidation value

   Whenever the project ends without being successful, a positive liquidation value is collected. This enhances the social value of the venture project and the financing horizon can, thus, be prolonged. If, however, the financial contract is a pure equity contract which gives part of the liquidation value to the entrepreneur, the incentive structure for a truthful allocation of funds is weakened. The optimal contract in face of a positive liquidation value should, therefore, use a mixture of debt and equity, e.g. convertible securities: The face value of debt amounts to the liquidation value, which, in case of successful innovation, will be converted into equity.

b) Monitoring

   If the venture capital company engages in monitoring activities, she obtains full information about the allocation of funds in the present period. Monitoring, however, is costly. The moral hazard problem is eliminated in the periods in which monitoring takes place, which might lead, in turn, to an extended financing horizon. The model of Bergemann and Hege suggests that monitoring is highest in later–stage periods.

Discussion

Bergemann and Hege (1998) investigate venture capital financing of an innovation project with uncertain returns in a multi-period context. The analysis focuses on a dynamic moral hazard problem between the entrepreneur and the venture capital company: The venture capitalist cannot observe how the entrepreneur uses the funds provided. In some cases, the entrepreneur may employ the invested capital for her own perquisites, which does not advance the project's discovery process. The authors' approach is an interesting and novel one, because the informational asymmetry and the agency problems in venture capital relationships are, for the first time, analyzed in a dynamical context. The paper examines the interaction between the investment and the learning process, which provides additional insights into the optimal financing of risky innovation.

Many results stand in accordance with empirical facts: First, the paper predicts that the entrepreneur's share of profits decreases over time. This is due, on the one hand, to the intertemporal incentive constraint and the high learning rent at the beginning of the project. On the other hand, it is due to the fact that the venture capitalist needs to be compensated for additional financing rounds and therefore obtains a higher share of profits if the project is successfully realized in a later

4.2 Venture capital financing – the individual firm's perspective

period. Empirical evidence shows that the entrepreneur's equity stake will be diluted if the project doesn't immediately succeed. In this case many shares have to be transferred to the venture capitalist to compensate her for the additional investment rounds.

Second, it is empirically evident that due to information problems in the development of early-stage projects, many projects are stopped too early. Especially under short-term contracting, an early-stopping problems exists. Bergemann and Hege show that a long-term contract can mitigate this early-stopping problem because it offers the possibility for intertemporal transfers. This relaxes the venture capitalist's participation constraint, such that the project's funding horizon is extended.

However, there are also some predictions of the model which seem to be at odds with reality. The major problem concerns the number of financing rounds for bad quality projects: Usually, low quality projects tend to be revealed and eliminated quite early. In Bergemann and Hege's model, however, it is the unsuccessful projects that receive most financing rounds. Therefore, the innovation process here is quite similar to a gambling game where coins are thrown into a slot machine until the winning combination appears, or until the amount wasted is higher than any expected future return. Thus, although venture capital companies commit to provide funds for several investment rounds, they specify ex-ante the financing horizon within which they want to cash out their investment. Typically, this investment horizon is three to five years, and not $T=52$ as proposed in Bergemann and Hege.

The second problem concerns the (conditional) success probability of the project: In their numerical example, Bergemann and Hege assume that the success probability of a good innovation project is around fifteen percent $(\theta_t = 0,15)$. On the other hand, the initial beliefs of financing a good quality project type are extremely optimistic $(\alpha_t = 0,95)$. According to the BVK statistics (2000), however, around 40 percent of the innovation projects are successful, such that the success probability θ_t should be assumed to be much higher. In contrast to this, it is very difficult for venture capital companies to identify good quality projects. Although venture capital companies spend a large amount of time and effort on due diligence, they cannot predict very well whether the project will turn out to be of good or bad quality. Thus, the initial beliefs α_t about the projects good quality might actually be lower. If the underlying parameters are modified in such a way, this will influence the development of beliefs and the project funding horizon.

Third, the innovation project stops immediately after the entrepreneur has achieved the innovative break-through. The final return Π is exogenous. However, if a firm successfully innovates, the new product will have to compete with existing products or it will open a new market and create additional demand. Yet, the market structure and the competitive environment in which the young firm operates is not taken into account.

Moreover, as Lerner (1998) states, the information flow about the firm's prospects is not distributed uniformly over time, but additional information is revealed only at some discrete points (when the results of the clinical trial emerge, when the prototype is successfully developed, after the product is introduced into a test-market). Lerner points out that the authors neglect the link between information generation and financing, in that a new round of equity financing is usually provided only after the next milestone has been attained.

Lerner (1998, 738) also criticizes that the monitoring activities of the venture capital company are quite unrealistic. While in the Bergemann and Hege model monitoring is highest towards the end of the innovation process, Lerner states that venture capitalists will control permanently the activities and financial statements of their portfolio firm. Or, as Berger and Udell (1998) emphasize, venture capital companies specialize in selecting high-risk projects in their early development stage. These private equity investors have an advantage over other financial intermediaries in investing in risky projects, but only because they specialize in monitoring projects from an early stage on.

Lerner's last remark concerns the staging of capital. In Bergemann and Hege, the staging of capital increases the probability of a successful outcome. However, empirical evidence suggests that the staging mechanism in venture capital contracting is designed rather to prevent inefficient project continuation.

4.2.2.5 Discussion

To summarize our findings about the allocation of property rights in venture capital contracting, we first have to state that the models are all very heterogeneous. Nevertheless, they provide many interesting insights into real-world aspects of venture capital contracting.

Berlöf (1994) identifies three exit channels for venture capital projects and examines various financial instruments. He shows that the highest social value is obtained by using convertible securities. In his model, however, there is no effort choice and no R&D investment to put the project into reality.

Casamatta (1999), by contrast, assumes that effort from both the entrepreneur and the venture capitalist is needed to successfully develop the innovation project. When the entrepreneur lacks experience, a financial partnership with a venture capitalist, who is investor and advisor, is desirable. In Casamatta's model, both parties jointly exert effort and make a financial contribution to the project. The sum of efforts, in turn, determines the success probability of the innovation project.

In Admati and Pfleiderer (1994), the entrepreneur needs external equity financing for two investment periods. The first round of funding is provided by an informed venture capitalist. In the second round, a less well-informed outside investor may participate in the project. If the quality is bad, the project will be stopped after the first period. The innovation success probability here depends on the state of nature.

Bergemann and Hege (1998), finally, investigate venture capital financing in a multi-period framework. Financing is provided in the form of equity. Moreover, Bergemann and Hege study the interaction between project uncertainty and the entrepreneur's unobservable allocation of funds. This provides additional insights into the financing of risky innovation projects. In order to obtain the innovative break-through of the project, the venture capital company finances the research and development costs in each period.

However, none of these papers takes the competitive environment of the young firm into account. In the next subsection we, therefore, analyze venture capital financing of an innovation project from the perspective of industrial organization. We explicitly formalize the market entry of the young, innovative firm and its subsequent price competition with incumbent firms.

4.3 Venture capital financing and product market competition

> *One of the main factors determining success or failure of a venture project is the strategic behavior of competitors. Competitors with better access to financial resources, supplier networks, marketing capacities, and brand loyalty of customers can easily exhaust a young firm's financial funds. Very often a young firm underestimates the predation strategies of well-established firms, that are determined to defend their market shares.*   Sonja Sidler (1997)

The competitive environment in which the new firm operates has been neglected so far in the theoretical papers on venture capital financing. To bridge this gap, the present analysis investigates the interaction between venture capital financing and product market competition. More precisely, we study how the market structure of an industry and the strategic behavior of competitors affects the financial contracting between the young firm and the venture capitalist.

If a young firm successfully innovates and introduces its new product into a market, it will typically face competition with existing firms. The product innovation, e.g. a new allergy test, a laser-based instead of manual operation method, can replace or at least partially substitute existing products. Incumbent firms of this industry will certainly react to the market entry of the new firm: either by softly accommodating entry, or by strategically engaging in competition and predation. Expected returns from the young firm's innovation project are, thus, determined by the competitive environment in the product market.

Our model is characterized by the following features:

(i) Market entry and product market competition: The model extends the Bergemann and Hege (1998) framework by an entry and competition game. We explicitly model product innovation and subsequent price competition between the new firm and an existing firm. Thus, we endogenize the venture project's return.

(ii) Innovation: The innovation project consists of a start-up stage and, potentially, a market expansion stage. In the start-up stage, the firm attempts to realize the product innovation and to introduce its new product to the market. The associated R&D expenditures depend linearly on the innovation success probability. The outcome of the innovation project is uncertain and can be a success or a failure. If entry is successful, the entrepreneur will proceed to the market expansion stage. She will again invest in R&D expenditures and to realize a process innovation and to acquire additional market shares.

(iii) In contrast to Bergemann and Hege's multi-period framework, we restrict ourselves to two periods of financing and competition. This is more realistic, because venture capital companies typically supply funds for two or three financing rounds at the maximum and invest in projects only for a previously specified, limited amount of time (Gompers and Lerner 1999c, chapter 7). On the other hand, our two-period framework still enables us to take intertemporal aspects into account.

(iv) Financial contracting: We analyze short-term and long-term financing. Under short-term contracting, the venture capital company supplies funds for just one period, and a different contract is written for each investment stage. Under long-term contracting, the venture capital company commits to finance the R&D expenditures in both periods.

(v) The principal-agent relationship between the young firm and the venture capital company is affected by a conflict of interest, i.e. moral hazard. Though the venture capitalist is perfectly informed about the profits in each period, she cannot observe the truthful investment of funds. The entrepreneur is, therefore, able to divert the funds to her private ends, and a problem of dynamic moral hazard arises.

(vi) Uncertain project quality: The innovation project's type is initially uncertain and more information arrives by developing the project. If the entrepreneur misuses the funds and the project consequently fails, this will affect the updating of beliefs about the project's quality. The entrepreneur, thus, gets access to an information rent, which imposes additional agency costs on the financial contract.

4.3 Venture capital financing and product market competition 153

The subsequent analysis is organized as follows: In subsection (4.3.1), we present the assumptions of our two-period innovation-, market entry- and competition game. In subsection (4.3.2), we assume that the firm does not have sufficient funds to self-finance the multi-period innovation project. The entrepreneur, therefore, needs external equity financing to cover her R&D expenditures. We analyze the provision of funds under short-term and long-term contracting. In subsection (4.3.3), we finally investigate the strategic reactions of the incumbent towards his new competitor. We are interested in how the competitive strategies of the existing firm will influence the financial contract between the young firm and the venture capital company.

4.3.1 Assumptions

A1: Players: Consider a market in which an incumbent enjoys monopoly profits.[7] An entrepreneur has an idea how to produce a similar product, in order to reap a share of the monopoly profits. If the innovation is successful, the monopoly is replaced by duopolistic price competition with heterogeneous products. The entrepreneur is wealth-constrained and seeks outside equity financing from a venture capitalist. The venture capital industry is characterized by perfect competition.

A2: The time horizon of the innovation project and the young firm's competition with the incumbent covers two periods. Each period consists of a research and development stage and a potential price competition stage.

A3: Information structure and learning:

1. Information about the project's type: Initially, there is symmetric non-information about the innovation project's type. Every agent believes that the project is either good with prior probability α_1 or bad with probability $1-\alpha_1$. If the project is good, then with probability θ_t the innovation is successfully realized. This probability θ_t is an increasing function of the R&D expenditures. We assume the research costs $r(\theta_t)$ to be linearly increasing in the success probability θ_t: $r(\theta_t)=g\theta_t$ with the cost parameter $g>0$ and the upper limit $g\theta^{max}$, above which additional funds do not increase the success probability of the project. We use this linear R&D technology for pure simplicity reasons (see Bergemann and Hege (1998) for a similar approach).

---

[7] Industrial organization is a research area with a large variety of competing models. As we have mentioned before, our approach is useful to study product innovation and entry of new firms in existing markets à la Schumpeter. We admit, though, that start-up firms from IT-business and e-commerce rather create new markets than enter into an existing ones. Innovation projects from these fields should, therefore, be analyzed in different settings, e.g. in terms of an uncertain market environment or in terms of network economics.

If the project is bad, then it always fails and yields zero return, independent of the amount of R&D expenditures invested.

2. Learning process: The uncertainty of the project's quality is (partially) resolved over time. In the start-up process, the young firm and the venture capitalist learn more about the prospects of the innovation project: If no success has occurred after the first period, all parties update their beliefs according to Bayes-rule: The posterior belief α_2 of developing a good project in the second period is equal to

$$\alpha_2 = \frac{\alpha_1(1-\theta_1)}{\alpha_1(1-\theta_1)+1-\alpha_1} \leq \alpha_1.$$

The revised beliefs α_2 are smaller than the initial beliefs α_1 (for $\theta_1 \neq 0$) (see also Appendix 4.5.2).

3. Asymmetric information about the investment of funds: The venture capital company is unable to observe whether the entrepreneur truthfully invests the funds or diverts them for private consumption. Here, a problem of moral hazard arises. The payoffs at the end of each period, however, are observable and verifiable by all parties, which is due to detailed financial statements required in venture capital contracting. The financial contract is, therefore, be based upon these per-period profits.

A4: Risk preferences: The entrepreneur and the incumbent are assumed to be risk-neutral.

A5: Production technology: To study market entry and competition, we use a well-known approach from industrial organization: The circular city of Salop (1979). The model consists of two steps: First, firms make their entry and location decisions. Second, firms engage in price competition with their heterogeneous products. Instead of simultaneous entry, however, we modify this game and consider sequential entry of the young firm after the monopolist has already established himself in the market.

If the young firm successfully enters, both firms will produce with the same marginal production costs c. In case the young firm reaches the expansion stage, the entrepreneur will attempt to innovate again in order to reduce the marginal production costs. The entrepreneur can, thus, lower the price, further expand in the market, and steal a part of the incumbent's business.

A6: Demand side: According to the Salop (1979) model, consumers are uniformly distributed along the circular city and all travel occurs around the circle. Consumers wish to buy one unit of the good per period of time. In addition to the product price, they bear disutilities T if the product variant of their choice is not supplied. In order to take dynamic aspects into account, we extend Salop's single-period framework to a model with two subsequent purchasing periods.

A7: Financial contract: The entrepreneur makes a take-it-or-leave-it offer to the venture capital company. The venture capitalist supplies equity capital to finance the R&D expenditures $g\theta_t$ in each period. In exchange for the capital provided, the venture capitalist obtains a share $1-S_t$ of the uncertain project's returns. These returns are observed by both parties at the end of each period. The entrepreneur and the venture capital company can commit to either short-term or long-term contracting. Due to perfect competition in teh venture capital market, the venture capitalist signs the financial contract as long as the expected returns are nonnegative.

A8: The contract does not specify any reallocation of control rights. Thus, it is always the entrepreneur who manages the project.

Market demand

If the young firm successfully enters, market demand will be divided between the two firms: Given the incumbent's location the young firm will choose to locate exactly opposite to him on the circle in order to attract customers whose preferences deviate most from the existing product.[8] Depending on the price p_i, the rival's price p_y, and the consumers' preferences T, firm i (i = incumbent) faces a demand of:

$$D_i(p_i, p_y, T) = \frac{p_y - p_i + T/2}{T}; \qquad (4.40)$$

while firm y (y = young firm) serves the rest of the market, $D_y(p_i, p_y, T) = 1 - D_i(p_i, p_y, T)$. Thus, the firm-specific demand increases the lower the own price and the higher the rival's price.

Profit possibilities

If the incumbent serves the entire market, his market share equals $D_i=1$. The monopoly profits Π^M are given by:

$$\Pi^M(p_i^M, c_i, T) = (p_i^M - c_i) \cdot 1 = \bar{s} - \tfrac{1}{2}T - c_i \qquad (4.41)$$

where $\bar{s}$ represents the consumers' reservation value and c_i the marginal production costs. For simplicity reasons and in order to obtain clear results for the

---

[8] Principle of maximal differentiation: The model shows that for profit maximizing reasons, firms locate equidistantly. The firms differentiate their products in order to soften price competition.

innovation and pricing strategies, we assume that the incumbent covers the entire market and does not think of covering only part of the market. If the monopolist maximized over the market demand to be served, q^M, instead, his profit function would be given by $\Pi^M = (\bar{s} - \frac{1}{2}Tq^M - c)q^M$, differentiation would yield $q^{M*} = \frac{1}{T}(\bar{s} - c)$, and the optimal profits would be given by $\Pi^M = \frac{1}{2T}(\bar{s} - c)^2$ (see Salop 1979, 143pp.). Thus, as long as $T \geq (\bar{s} - c)$, it follows that the total market is served $q^{M*} \geq 1$.

If the young firm fails to innovate and cannot enter the market, it will realize gross profits of zero. In this case, the incumbent continues to earn monopoly profits. If the young firm successfully enters, both firms will produce with constant marginal costs c. Fixed production costs are assumed to be zero. The profit function for firm i equals (see Appendix 4.5.1):

$$\Pi_i(p_i, p_y, c_i, T) = (p_i - c_i)D_i = (p_i - c_i)(\frac{p_y - p_i + T/2}{T}), \qquad (4.42)$$

and for firm y, analogously. Both firms choose their optimal prices such as to maximize their per-period profits, $p_i^* = p_y^*$. In this case, market shares equal one half and profits are

$$\Pi_i(T) = \Pi_y(T) = T/4. \qquad (4.43)$$

We see that these profits depend on the size of the consumers' preferences, T, but not on the marginal costs, c. If, however, the young firm innovates again in the second-period expansion stage, it will realize a cost advantage. We express this cost difference by $\Delta c = c_i - c_y$. This cost-advantage enables the young firm to reduce its product price and to attract a higher market share, i.e. $D_y^A > D_i^D$. Its profits (superscript A for cost-advantage, superscript D for disadvantage) are then given as:

$$\Pi^A(\Delta c, T) = T/4 + \Delta c/3 + (\Delta c/3)^2/T. \qquad (4.44)$$

These profits depend positively on the cost-difference Δc. At the same time, the incumbent realizes profits of:

$$\Pi^D(\Delta c, T) = T/4 - \Delta c/3 + (\Delta c/3)^2/T. \qquad (4.45)$$

These profits are negatively influenced by the cost-difference. Independent of the cost-situation, however, the young firm realizes higher profits when consumers' preferences T are strong: The intuition behind this is that the new product variant is more suitable to many customers.

4.3.1.1 The basic model

The basic game in our model consists of two steps: In the first stage, the young firm invests in R&D in order to realize a product innovation. Nature decides whether the outcome of the project is successful or fails. If it is a success, then in the second stage, both firms choose their duopoly prices, sell their product variants and realize the respective profits. If the project fails, the young firm cannot enter and the existing firm continues to realize its monopoly profits. The time-line of this basic innovation, entry and competition game is as follows:

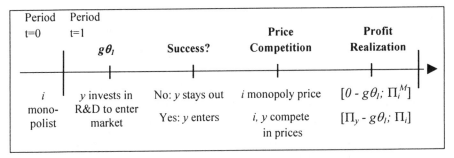

Figure 4.4: Time-line of the basic entry game

We suppose that the incumbent has no way to deter the entry of the young firm.[9] We solve this game by backward induction: Both firms choose their optimal prices, given that the young firm has entered the market. Since both firms produce with the same marginal costs, they will charge the same optimal product price.[10] The market splits into two equal shares, and the firms realize symmetric profits, $\Pi_i = \Pi_y$. If entry fails, the young firm realizes a gross return of zero. Given these prices and profit levels, the young firm determines its optimal R&D expenditures. The success probability of market entry is $\alpha_1 \theta_1$, which depends on both the estimated quality of the project and, indirectly, on the level of R&D expenditures $g\theta_1$. The failure probability is given by $1 - \alpha_1 \theta_1$. Therefore, the expected value of the innovation project equals:

$$V(\theta_1) = \alpha_1 \theta_1 \Pi_y - g\theta_1. \tag{4.46}$$

---

[9] We will modify this assumption in section (4.3.3) below.
[10] In Milgrom and Roberts's (1982) limit pricing model, the entrant's marginal costs c can be higher or lower, such that the young firm supplies either at a higher or lower product price. In Milgrom and Roberts's approach, however, consumer's preferences are identical (no transportation costs), price competition is in homogeneous goods which results in that only one firm (entrant or incumbent) finally supplies to the market. In contrast to this, our model assumes a symmetric market position after the young firm has entered for the first time.

The firm maximizes the project value over the success probability θ_l:

$$\partial V(\theta_1)/\partial \theta_1 = \alpha_1 \Pi_y - g \ . \tag{4.47}$$

Since the project value is linear in θ_l, the firm will invest the maximal amount $g\theta_1^{max}$ in R&D activities as long as $\alpha_1 \Pi_y \geq g$ (or as long as $\alpha_1[\frac{1}{4}T] > g$). For $\alpha_1 \Pi_y < g$, however, the firm will spend nothing on R&D activities at all, such that $g\theta_l = 0$.

4.3.1.2 Innovation, entry, and competition in the two-period framework

We now extend our model to two periods of innovation and competition. The idea is that when the young firm failed to enter in the first period, it will attempt to yield a breakthrough in the subsequent period and to again challenge the monopolist. Thus, the firm has another opportunity to spend on R&D activities and to innovate. If however the young firm has already entered the market at first place, it will - in contrast to Bergemann and Hege (1998) - continue to develop its product: In the second period, it spends again on R&D activities in order to reduce the marginal production costs (or to improve the product's quality) and to gain a competitive advantage over the incumbent. In this manner, the young firm intends to lower its product price, to further expand into the market and to increase its profit level. By extending the time horizon to a second period of innovation and either entry or repeated competition, our model combines product innovation with subsequent process innovation. Empirically, this phenomenon is often observed in the life-cycle of technology-intensive products (see e.g. Pfirrmann, Wupperfeld, and Lerner, 1997, on the development of new firms in the telecommunication sector, the pharmaceutical industry, or the laser optical equipment market). The extended form of this two-period innovation-, entry- and competition game is as follows:

4.3 Venture capital financing and product market competition

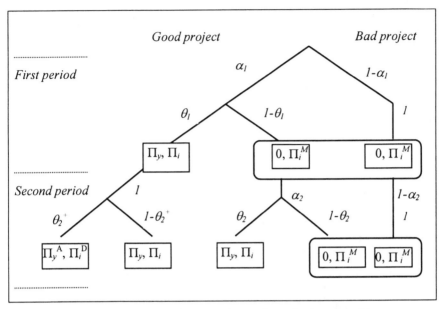

Figure 4.5: The two-period game of innovation and price competition

We see that at the end of period two, some projects (the left-hand side subgame in the second period) are well established in the market and become true shooting stars, some projects achieve market entry only in the second period and obtain a mediocre return, whereas the remaining fraction of projects fails completely and earns zero gross profits. In that case, the entrepreneur does not know whether the failure is due to the fact that a good project didn't succeed twice or whether she invested into a bad project. Thus, the uncertainty about the project's quality is not completely resolved at the end of the time horizon. The distribution of profits in our model corresponds to the stylized fact that returns of high-risk innovation projects vary widely, ranging from payoffs many times greater than investment costs to projects experiencing a total loss (Sahlman 1990, BVK 2001).

Given our potential profit levels from the second-period price competition, we can derive the optimal R&D expenditures for each subgame. Starting with the second period expansion stage, the firm maximizes the value of the innovation project in this subgame as:

$$\max_{\theta_2^+} V(\theta_2^+) = \theta_2^+ \Pi_y^A + (1-\theta_2^+)\Pi_y - g\theta_2^+, \qquad (4.48)$$

where θ_2^+ denotes the success probability of the process innovation. Since (4.48) is linear in θ_2^+, we see that for $\Pi_y^A - \Pi_y \geq g$, the firm invests the maximum amount of R&D expenditures, such that $\theta_2^+ = \theta^{max}$; whereas for $\Pi_y^A - \Pi_y < g$, the firm does not invest anything at all, such that $\theta_2^+ = 0$.

In the other second-period subgame, i.e. the second attempt of market entry, the firm maximizes the expected project value according to:

$$\max_{\theta_2} V(\theta_2) = \alpha_2 \theta_2 \Pi_y - g\theta_2. \tag{4.49}$$

Again, since (4.49) is a linear function in the success probability θ_2, the firm will invest the maximal amount of R&D expenditures, $g\theta_2 = g\theta^{max}$, as long as $\alpha_2 \Pi_y \geq g$ holds, and nothing $(\theta_2 = 0)$ otherwise.

Stepping back to the first period of innovation, we start by calculating the total expected value of the venture. Here, we have to take into account that the expected value of the expansion stage is conditioned on the probability of market entry in the first period, $\alpha_1 \theta_1$; and that the expected value of the second-period entry stage depends on the probability of failure in the first period, $1 - \alpha_1 \theta_1$. The total expected value of the innovation project is (no discounting):

$$\begin{aligned} V(\theta_1, \theta_2, \theta_2^+) = & \ \alpha_1 \theta_1 \Pi_y - g\theta_1 \\ & + \alpha_1 \theta_1 [\theta_2^+ \Pi_y^A + (1 - \theta_2^+) \Pi_y - g\theta_2^+] \\ & + (1 - \alpha_1 \theta_1)[\alpha_2 \theta_2 \Pi_y - g\theta_2]. \end{aligned} \tag{4.50}$$

It consists of the expected net return in the first period, plus the expected returns if the firm tries to capture additional market shares in the second period of competition, plus the returns if entry occurs only after the second R&D investment round. Rearranging terms, the total value can be expressed as:

$$\begin{aligned} V(\theta_1, \theta_2, \theta_2^+) = & \ \alpha_1 \sum_{t=1}^{2} (\theta_t \Pi_y - g\theta_t)(1 - \theta_1)^{t-1} \\ & - (1 - \alpha_1) \sum_{t=1}^{2} g\theta_t \\ & + \alpha_1 \theta_1 [\theta_2^+ \Pi_y^A + (1 - \theta_2^+) \Pi_y - g\theta_2^+]. \end{aligned} \tag{4.51}$$

Here, we see that the total expected value of the project is the sum of net returns if entry occurs in the first or in the second period, reduced by the sum of R&D investments wasted on bad projects, and increased by additional returns if the firm further develops its product in the expansion phase.

4.3 Venture capital financing and product market competition

How does the expected value of the venture project depend on the R&D investment in the first period? Taking the derivative of (4.50) and replacing α_2 by the Bayes-formula, we obtain:

$$\frac{\partial V(\theta_1, \theta_2, \theta_2^+)}{\partial \theta_1} = \alpha_1 \Pi_y - g + \underbrace{\alpha_1 [\theta_2^+ \Pi_y^A + (1-\theta_2^+)\Pi_y - \theta_2 \Pi_y]}_{>0}. \qquad (4.52)$$

The value of the venture project is again linear in θ_1 and, therefore, it is optimal to either allocate the maximum amount of capital or not to allocate any venture capital at all. Thus, if the marginal expected return exceeds the marginal research costs in equation (4.52), the firm will invest the maximal amount in R&D expenditures, such that $\theta_1 = \theta^{\max}$. Due to the additional term on the right hand side, expression (4.52) is less restrictive than condition (4.46) from the single-period case. Moreover, because $\alpha_2 < \alpha_1$, expression (4.52) is also less restrictive than the investment condition from the second-period market entry stage (4.49). Therefore, we can state:

Proposition 4.10 (Investment Policy)

If the project is profitable (i.e. the investment conditions $\alpha_2 \Pi_y > g$ and $\Pi_y^A - \Pi_y > g$ are fulfilled), it is optimal to invest the maximum amount of capital in each period into the R&D project, such that $\theta_t = \theta^{\max}$ for t=1,2.

4.3.2 Venture capital contracting

In this subsection, we suppose that the young firm does not have sufficient funds to develop its product. Moreover, the firm is subject to limited liability. In order to circumvent the financial restrictions, the entrepreneur seeks equity financing from the venture capital company. High-growth firms which are characterized by significant intangible assets and tremendous market uncertainties are unlikely to receive debt financing, which makes venture capital the only source of funding in early stages (see Fenn and Liang, 1998; Gompers and Lerner, 1999).

In exchange for the provision of capital, the venture capital company obtains a share of the risky, but potentially high returns from the project. The venture capital industry is characterized by perfect competition. This reflects the fact that there is abundant capital available seeking profitable investment (Reid, 1996; Gompers, 1998). The young firm suggests financial contracts to any of the venture capital companies. The bargaining power is on the side of the firm. The selected venture capitalist then decides whether to accept or reject the financial contract. A venture capital company agrees to finance the project if it receives nonnegative expected profits.[11]

Venture capital companies usually spend considerable time to screen and evaluate potential projects. We suppose, however, that the venture capital company has gone through the whole due diligence process and believes with probability α_l that the project is of good quality. The entrepreneur shares the same initial beliefs about the project's quality, such that at the beginning there is symmetric non-information. More information arrives by developing the project. Berger and Udell (1998) point out that informational opaqueness is extremely acute in young start-up firms.

This implies in turn, that even though the financial relationship between the venture capital company and the young firm is very close, entrepreneurs still can acquire private information about the projects they develop: Venture capitalists receive regular financial reports and have monthly meetings with their portfolio company, but they typically do not become involved in the day-to-day business of the firm (Gompers, 1995).

Thus, the entrepreneur may behave opportunistically and spend insufficient effort on the project or exhibit expense preference behavior. This leads to a problem of moral hazard with hidden action. Upon receiving the venture funds, the entrepreneur can either allocate the funds truthfully towards the development of the product, or she can misuse the funds and divert them to her private ends. The venture capital company is unable to observe the proper investment of funds, because e.g. high monitoring costs prevent a detailed understanding of the technical position of the young firm (see also Gompers and Lerner, 1996). We assume, however, that the venture capital company is informed about *(i)* the level of gross profits realized at the end of each financing period $[0, \Pi_y, \Pi_y^A]$, *(ii)* the number of firms operating in the industry (monopoly or duopoly), and *(iii)* how many times the young firm has attempted to innovate.[12]

4.3.2.1 Short-term contracting

We begin by analyzing the procurement of venture capital under short-term financing. To correspond to the information available at each stage, the firm offers a different financial contract for each scenario: In the first period, the contract states that the firm obtains the share S_l of the expected profits, whereas the venture capital company obtains the remaining share, $(1-S_l)$. In the second period,

---

[11] As in Diamond's (1984) model of delegated monitoring, the depositors who provide the original funds for the venture capital company require a net return of R at the minimum (see also Ramakrishnan and Thakor (1984)). In our model, we assume that this return R equals zero.

[12] We assume that after each financing period, the firm receives a time-label for identification.

the firm obtains S_2 in case of second-period entry, and S_2^{Duo} in the expansion stage, while the venture capital company receives the remaining shares, $(1-S_2)$ and $(1-S_2^{Duo})$, respectively. Thus, the optimal short-term contracts are simple share contracts between the young firm and the venture capitalist.

First-period contracting

We start with the financial contract for the first-period market entry stage. The young firm's objective function is to maximize the expected value of its share of profits:

$$\max_{\theta_1, S_1} V_{t=1}^{ST} = S_1 \alpha_1 \theta_1 \Pi_y. \tag{4.53}$$

Here, the superscript ST stands for short-term financing. In the case of external financing, the young firm does not make any monetary investments into the innovation project. However, the firm must have sufficient incentives to truthfully invest the funds. The incentive compatibility constraint (IC) for the young firm is given by:

$$\alpha_1 \theta_1 S_1 \Pi_y - g\theta_1 \geq 0. \qquad \text{(IC)} \tag{4.54}$$

Therefore, the entrepreneur's expected profits have to exceed the amount of research expenditures invested in this period. Rearranging terms, we solve for the minimum share of profits that the entrepreneur requires for truthful investment in the first period:

$$S_1 \geq \frac{g}{\alpha_1 \Pi_y}. \tag{4.55}$$

We see that for $\theta_1 \neq 0$, this minimum share is independent of the innovation success probability, which is due to the fact that the R&D cost function is linear.

The venture capital company, on the other hand, decides whether or not to participate in the financial contract upon the following restriction:

$$(1-S_1)\alpha_1 \theta_1 \Pi_y - g\theta_1 \geq 0. \qquad \text{(PC)} \tag{4.56}$$

The venture capital company is willing to accept the present-period's financial engagement, if its expected profits exceed the amount of funds invested. Since all the bargaining power is on the side of the firm, the venture capital company's expected profits are reduced to zero, and the participation constraint in (4.56) is binding. Rearranging terms, we see that the minimum share of expected profits that the venture capitalist requires in $t=1$ for participating in the financial contract, is:

$$(1-S_1) = \frac{g}{\alpha_1 \Pi_y} \quad . \tag{4.57}$$

The young firm maximizes the value of its expected share of profits (4.53) under the incentive constraint (4.54) and the participation constraint of the venture capital company (4.56). The financial contract is conditioned on the level of profits (Π_y in case of success, and 0 in case of failure) as well as on the respective shares of the contracting parties, S_1 and $(1-S_1)$.

Adding up the two constraints (4.54), (4.56), we see that the expected profits from the venture project must be at least twice as high as the R&D expenditures

$$\alpha_1 \theta_1 \Pi_y \geq 2 g \theta_1 \tag{4.58}$$

in order for the firm to obtain financing in the first period.[13] Thus, the more expensive the R&D costs $g\theta_1$, or the lower the beliefs α_1 about the project's quality, the higher must be the profit level after market entry to obtain initial financing. From subsection 3.2.1 we know that profits are high if consumers have strong preferences T for the product variants supplied. On the other hand, the success probability of the good project has no influence on the minimum threshold profit level as long as the R&D cost function is linear.

Second-period market entry stage

Proceeding to the second period market entry stage, i.e. the case in which the firm has failed to innovate in the first period but attempts, again, to enter in the second period, the firm's maximization problem has exactly the same structure as above. The firm maximizes

$$\max_{\theta_2, S_2} V_{t=2}^{ST} = S_2 \alpha_2 \theta_2 \Pi_y$$

s.t. (IC) $\quad S_2 \alpha_2 \theta_2 \Pi_y - g\theta_2 \geq 0,$ (4.59)

(PC) $\quad (1-S_2)\alpha_2 \theta_2 \Pi_y - g\theta_2 \geq 0.$

Since the updated beliefs α_2 are smaller than the first-period beliefs α_1, the incentive and participation constraints in (4.59) become more restrictive than under the first-period problem. Adding up the two second-period constraints (IC)

---

[13] The competing claims emanating from the investment problem of the venture capital company and the agency problem of the entrepreneur lead to a conflict of interest if profits cannot cover the total remuneration for both the participation and incentive compatibility constraints. The financing of the venture project, thus, resembles a team problem à la Holmström (1982), where two parties contribute and the budget is unbalanced.

and (PC) of (4.59), the profit level for the project to obtain financing for second period market entry must be:

$$\alpha_2 \theta_2 \Pi_y \geq 2g\theta_2. \tag{4.60}$$

This minimum threshold of profits in the second period must be higher than in the first period due to $\alpha_2 < \alpha_1$. It is intuitively clear that if a project fails once, it will be harder to obtain financing for this project in the subsequent period. Therefore, the share of profits accruing to the venture capital company ($1-S_2$) has to be larger than under first-period contracting.

Second-period expansion stage

Finally, if the firm needs external financing for the expansion stage,[14] the firm's maximization problem and the respective incentive compatibility and participation constraints in this subgame are given by:

$$\max_{\theta_2, S_2^{Duo}} V_{t=2Duo}^{ST} = S_2^{Duo}[\theta_2^+ \Pi_y^A + (1-\theta_2^+)\Pi_y]$$

$$\text{s.t.} \quad \text{(IC)} \quad S_2^{Duo}[\theta_2^+ \Pi_y^A + (1-\theta_2^+)\Pi_y] \geq g\theta_2^+ + S_2^{Duo}\Pi_y \tag{4.61}$$

$$\text{(PC)} \quad (1-S_2^{Duo})[\theta_2^+ \Pi_y^A + (1-\theta_2^+)\Pi_y] - g\theta_2^+ \geq 0.$$

Since the venture capital company is able to distinguish between the two profit levels realized, Π_y and Π_y^A, the financial contract can be conditioned upon these profit levels. The incentive constraint in this subgame states that the share of profits accruing to the firm under truthful investment must be higher than the sum of diverted research expenditures plus the share of sure profits that the entrepreneur obtains if no innovation takes place in the expansion stage. Rearranging terms leads to:

$$S_2^{Duo} \geq \frac{g}{(\Pi_y^A - \Pi_y)}. \tag{4.62}$$

The venture capitalist participates as long as the expected profits exceed the investment costs. Rearranging the participation constraint leads to:

$$(1-S_2^{Duo}) \geq \frac{g\theta_2^+}{\theta_2^+ \Pi_y^A + (1-\theta_2^+)\Pi_y}. \tag{4.63}$$

---

[14] If the profits of the first period are high enough such that the firm could self-finance the subsequent innovation and market expansion activities, the firm will not offer a financial contract for the second-period expansion phase.

Adding up both constraints, we derive the required minimum profit level that the project must attain in the expansion stage:

$$\theta_2^+ \Pi_y^A + (1-\theta_2^+)\Pi_y - S_2^{Duo}\Pi_y - 2g\theta_2^+ \geq 0. \qquad (4.64)$$

The expected profits of the expansion stage, less the share of symmetric profits accruing to the firm for incentive reasons, must be higher than twice the amount of R&D expenditures such that financing is obtained and funds are invested truthfully. We summarize our analysis of short-term contracting in the following proposition.

Proposition 4.11 (Short-term contracting)

Under short-term contracting, the firm writes a different financial contract Γ_t^{ST}, $t=1,2,2^+$ for each stage. The financial contracts for the first- and second-period market entry stages and the second period expansion stage, conditional on the fact that a minimum threshold of profits is attained, are given by:

$$\Gamma_{t=1}^{ST}\left[S_1,(1-S_1)\big|\Pi_y^{1\min}\right],$$

$$\text{with } S_1 \geq \frac{g}{\alpha_1 \Pi_y}; \quad (1-S_1) = \frac{g}{\alpha_1 \Pi_y}; \quad \Pi_y^{1\min} = \frac{2g}{\alpha_1}, \quad \theta_1, \alpha_1 \neq 0.$$

$$\Gamma_{t=2}^{ST}\left[S_2,(1-S_2)\big|\Pi_y^{2\min}\right],$$

$$\text{with } S_2 \geq \frac{g}{\alpha_2 \Pi_y}; \quad (1-S_2) = \frac{g}{\alpha_2 \Pi_y}; \quad \Pi_y^{2\min} = \frac{2g}{\alpha_2}, \quad \theta_2, \alpha_2 \neq 0.$$

$$\Gamma_{t=2Duo}^{ST}\left[S_2^{Duo},(1-S_2^{Duo})\big|\Pi_y^A,\Pi_y,\theta_2^+\right],$$

$$\text{with } S_2^{Duo} \geq \frac{g}{\Pi_y^A - \Pi_y}; \quad (1-S_2^{Duo}) = \frac{g\theta_2^+}{\theta_2^+ \Pi_y^A + (1-\theta_2^+)\Pi_y};$$

$$\theta_2^+ \Pi_y^A + (1-\theta_2^+)\Pi_y - S_2^{Duo}\Pi_y - 2g\theta_2^+ \geq 0.$$

We state again that in each financial contract the venture capital company obtains no more than the minimum share of profits that it requires for participation in the contract. The firm, on the other hand, obtains at least the minimum share of profits necessary for incentive reasons, as well as all additional profits in case of financial slack, i.e. if the project's profits surpass the minimum threshold level.

Comparison of the short-term contracts

In addition this, we want to know in which stage it is most difficult to obtain short-term financing for the young firm. Therefore, we compare the total project financing restrictions of the first and second period market entry stages, (4.58) and (4.60), with the restriction of the expansion stage (4.64):

$$\alpha_1 \theta_1 \Pi_y - 2g\theta_1 \geq 0, \tag{4.58}$$

$$\alpha_2 \theta_2 \Pi_y - 2g\theta_2 \geq 0, \tag{4.60}$$

$$\theta_2^+ \Pi_y^A + (1-\theta_2^+)\Pi_y - S_2^{Duo}\Pi_y - 2g\theta_2^+ \geq 0. \tag{4.64}$$

As we have stated above, due to $\alpha_2 < \alpha_1$, early-stage financing is harder to obtain in the second than in the first period. Thus, condition (4.60) of the second-period entry stage is more restrictive than its counterpart (4.58) from the first period.

Next, we compare the market entry stage restriction with the expansion stage condition (4.64). Financing is easier to obtain in the expansion stage if the following holds:

$$\theta_2^+ \Pi_y^A + (1-\theta_2^+)\Pi_y - S_2^{Duo}\Pi_y \geq \alpha_1 \theta_1 \Pi_y, \tag{4.65}$$

where $\theta_1 = \theta_2' = \theta^{max}$. After rearranging terms, we can state:

Proposition 4.12 (Financial restrictions)

The ranking of the financial contracts shows that for $[\Pi_y^A/\Pi_y \geq 1 + 2\alpha_1 - (1-S_2^{Duo})/\theta^{max}]$:[15]

1. *Short-term financing is easier to obtain in the expansion phase (after market entry) than in the startup stage (before market entry).*

2. *Short-term financing is most difficult to obtain for the second-period start-up phase after the young firm has already failed once to enter the market.*

Proposition 4.12 stands in accordance with the empirical fact that many venture capital companies prefer to finance the expansion stage of young firms, whereas only few venture capitalists specialize in startup financing. In Germany, only 15

---

[15] In case that the inequality does not hold, the ranking of financial contracts concerning the availability of funds is altered to {first entry ≻ expansion stage ≻ second entry}, or even to {first entry ≻ second entry ≻ expansion stage}.

per cent of venture funds are allocated to seed and start-up financing, while expansion-stage financing accounts for 55 per cent of the investments (BVK 2000). Similarly, the OECD reports for Europe that the lion's share of venture capital investments are dedicated to later stage investments in established businesses and management buyouts/buyins (Organisation of Economic Cooperation and Development, 1996). The US venture capital industry, however, has shifted from 15 per cent start-up financing, 65 per cent later-stage investment, and 20 per cent leveraged buyouts deals during the 1980s (Sahlman, 1990) towards a higher investment into early-stage businesses. Nowadays, US venture capital companies invest in almost as many early-stage as in later-stage companies, which is due, as Black and Gilson (1998) state, to the more complete development of the US venture capital market.

However, since short-term contracts do not allow for intertemporal transfers, they typically restrict financing to projects with very high expected returns. In the next subsection we therefore analyze whether long-term contracting will improve the financing possibilities for the young firm.

4.3.2.2 Long-term contracting

Empirically, the financing horizon of venture capital projects is rather medium or long-term than short-term: Venture capital companies typically provide funds into a single company for a period of three to five years (Gompers, 1998). We investigate now how the commitment of the venture capitalist to supply funds for two subsequent periods will change the financial contracting. Thus, the various short-term contracts are replaced by a single long-term contract, $\Gamma^{LT}(S_1^{LT}, S_2^{LT}, S_{2Duo}^{LT}, \theta_1, \theta_2, \theta_2^+)$, which covers the whole project horizon. The financial contract is a time-varying share contract which specifies both parties' shares according to the relevant scenario in each period.[16]

Moreover, the long-term financial contract provides room for intertemporal transfers: Under short-term financing, the young firm will not be refinanced if expected profits in the second market entry stage are low, i.e. are drawn from the interval $[2g\theta_2 > \alpha_2\theta_2\Pi_y \geq g\theta_2]$. Under self-financing, however, the project would be profitable. Here, a long-term contract helps to mitigate the early-stopping problem: The venture capital company, instead of breaking even in each period, accepts to invest in the project as long as the expected repayments cover the R&D costs of both periods. The sequence of participation constraints is, thus, replaced by a single intertemporal participation constraint.

---

[16] This stands in contrast to the results of Admati and Pfleiderer (1994), who show that a time-invariant share contract is optimal for venture projects under uncertainty.

4.3 Venture capital financing and product market competition

If the venture capital company agrees to the long-term financial contract, second-period refinancing is guaranteed with certainty. This means that if market entry fails in the first period, the firm will have another chance to innovate and to introduce its product to the market.

At the same time, a long-term contract provides more opportunities to divert funds, due to the intertemporal structure of the financial arrangement. Firstly, the firm can divert the provided funds in either investment periods. Secondly, the firm can divert the funds today and can bet on a positive realization of the innovation project tomorrow. In this case, the learning process about the project's quality becomes asymmetric: The venture capital company, on the one hand, updates its beliefs because the project's return after the first period is zero. The entrepreneur, on the other hand, knows that the project failed as a consequence of her hidden action decision and, therefore, does not adjust her beliefs about the project's quality. Information becomes asymmetric, and the entrepreneur is granted an information rent (to be specified below).

If the funds are invested truthfully and if the venture capital company agrees to finance the project, the maximization problem of the entrepreneur under long-term contracting is given as follows:

$$\max_{\substack{\theta_1, \theta_2 \\ S_1^{LT}, S_2^{LT}, S_{2Duo}^{LT}}} V^{LT} = S_1^{LT} \alpha_1 \theta_1 \Pi_y$$

$$+ \alpha_1 \theta_1 S_{2Duo}^{LT} [\theta_2^+ \Pi_y^A + (1-\theta_2^+) \Pi_y] \quad (4.66)$$

$$+ (1-\alpha_1 \theta_1)[\alpha_2 \theta_2 S_2^{LT} \Pi_y].$$

We state again that the firm does not make any monetary contribution during the entire project horizon. Equation (4.66) represents the total expected value of the project accruing to the entrepreneur, if she is willing to allocate the funds truthfully. Since under long-term contracting, the entrepreneur has different possibilities to divert the funds during the whole financing horizon, we have to account for four different incentive compatibility conditions, (4.67 a-d):

$$S_{2Duo}^{LT}[\theta_2^+ (\Pi_y^A - \Pi_y) + \Pi_y] \geq g\theta_2^+ + S_{2Duo}^{LT} \Pi_y. \quad \text{(IC)} \quad (4.67a)$$

Condition (4.67a) will prevent the firm from misusing the funds in the second period if it has truthfully invested them in the first period and market entry has occurred.

$$\alpha_2 S_2^{LT} \theta_2 \Pi_y \geq g\theta_2. \quad \text{(IC)} \quad (4.67b)$$

Condition (4.67b) will prevent the firm from the misuse of funds in the second period if the firm has truthfully invested in the first period, but no entry has occurred.

$$V^{LT} \geq g\theta_1 + S_2^{LT}\alpha_1\theta_2\Pi_y.$$ (IC) (4.67c)

Condition (4.67c) states that the expected value under truthful investment should be higher than diverting first-period funds and making a truthful investment in the second period. Note that the posterior beliefs on the right-hand-side of this equation are α_1 instead of α_2 because the entrepreneur does not update the beliefs after deviation.

$$V^{LT} \geq g\theta_1 + g\theta_2.$$ (IC) (4.67d)

The incentive condition (4.67d) shall prevent the entrepreneur from misusing the funds in both periods.

Finally, the intertemporal participation constraint for the venture capital company is given as follows:

$$(1-S_1^{LT})\alpha_1\theta_1\Pi_y + \alpha_1\theta_1(1-S_{2Duo}^{LT})[\theta_2^+\Pi_y^A + (1-\theta_2^+)\Pi_y]$$
$$+ (1-S_2^{LT})\alpha_2\theta_2\Pi_y(1-\alpha_1\theta_1) \geq 2g\theta.$$ (PC) (4.68)

It states that the shares of profits that the venture capital company obtains in both periods must be larger than the sum of funds invested in the two financing periods.

We analyze the maximization problem in two steps: First we look for the minimal incentive compatible shares for which the entrepreneur is willing to allocate the funds truthfully over the whole financing horizon. Then, in a second step, we check under which conditions the participation constraint of the venture capital company (4.68) is fulfilled.

The intertemporal incentive problem

We solve the entrepreneur's incentive problem firm by backward induction. By looking for the minimum incentive-compatible shares, we know that the second-period conditions (4.67a) and (4.67b) must be binding:

$$S_{2Duo}^{LT} = g/(\Pi_y^A - \Pi_y),$$ (4.67a')
$$S_2^{LT} = g/\alpha_2\Pi_y.$$ (4.67b')

Next, by inserting the above expression (4.67b') into condition (4.67c), we get:

$$V^{LT} \geq g\theta_1 + [g\theta_2]\cdot(\alpha_1/\alpha_2).$$ (4.67c')

It states that the expected value under truthful long-term investment must be greater than first-period deviation plus second-period truthful allocation of funds. By comparing it to the last incentive condition (4.67d), we see that condition

4.3 Venture capital financing and product market competition   171

(4.67c') is more restrictive due to the term $(\alpha_1/\alpha_2) \geq 1$. This captures the fact, that under long-term financing, in addition to the sum of funds provided, $g\theta_1 + g\theta_2$, the entrepreneur gets access to an information rent. Thus, to prevent the entrepreneur from moral hazard, the share of profits under truthful investment (4.66) must be higher than or equal to the payoffs under deviation (4.67c'):

$$\alpha_1\theta_1 S_1^{LT}\Pi_y + \alpha_1\theta_1 S_{2Duo}^{LT} E(\Pi_{t=2}) + (1-\alpha_1\theta_1)(\alpha_2\theta_2 S_2^{LT}\Pi_y)$$
$$\geq g\theta_1 + \frac{\alpha_1}{\alpha_2}(\alpha_2\theta_2 S_2^{LT}\Pi_y). \tag{4.69}$$

The solution to this intertemporal incentive problem is summarized as follows:

Proposition 4.13 (Share contract)

The minimum shares of profits that the entrepreneur requires for truthful investment under a long-term contract are given by $S_1^{LT}, S_2^{LT}, S_{2Duo}^{LT}$, with

$$S_1^{LT} = \underbrace{\frac{g}{\alpha_1\Pi_y}}_{I} + \underbrace{\frac{g\theta_2}{\Pi_y}}_{II} - \underbrace{S_{2Duo}^{LT}\frac{E(\Pi_{t=2})}{\Pi_y}}_{III} + \underbrace{\left(\frac{\alpha_1}{\alpha_2} - \alpha_1\right)\frac{g\theta_2}{\alpha_1\Pi_y}}_{IV}, \tag{4.70}$$

$S_2^{LT} = g/\alpha_2\Pi_y$, and

$S_{2Duo}^{LT} = g/(\Pi_y^A - \Pi_y)$.

Proof. See Appendix 4.5.3.

The minimum share S_1^{LT} incorporates the intertemporal aspects of the financial contract. It ensures that the firm employs the capital in each period towards the discovery process or the improvement of the product. Equation (4.70) may seem rather inaccessible at first, but can be decomposed into different aspects of the agency problem between the entrepreneur and the venture capitalist:

I) Static agency costs: We see that term *I* of (4.70) is identical to the static incentive compatibility condition from short-term financing in the first period, (4.54),

$$\frac{g}{\alpha_1\Pi_y} \equiv S_1. \tag{4.54'}$$

This would be the minimum share for the entrepreneur if the innovation project was financed only during a single period.

II) Intertemporal agency costs: If the venture capital company agrees to long-term financing, but could observe the development of beliefs, such that the entrepreneur would not get access to the information rent, the minimum share (4.54') would have to be modified in two ways: On the one hand, the minimum share has to be increased by the second term *II*, which reflects the option for the entrepreneur to withhold financing for a single period, but to switch to truthful investment in the next period:

$$+\frac{g\theta_2}{\Pi_y}.$$

III) Competition effect: On the other hand, the minimum share has to be reduced by the share of profits the entrepreneur could have gained in the expansion stage (term *III*). Thus, the opportunity to increase potential profits in the second period of competition - if funds are employed appropriately in the first period - helps to realign the incentives of the entrepreneur:

$$-S_{2Duo}^{LT}\frac{E(\Pi_{t=2})}{\Pi_y}.$$

We denote the term above as the "competition effect". Moreover, we see that the effects *II* and *III* work in opposite directions: The withholding option, $g\theta_2/\Pi_y$, increases the minimum share of profits that must be given to the entrepreneur. By contrast, the competition effect reduces it. This negative effect of the foregone opportunities to gain additional profits in the expansion stage may even dominate the positive effect of the withholding option!

IV) Information rent: The informational agency costs to be added are expressed by term *IV*:

$$+\left(\frac{\alpha_1}{\alpha_2}-\alpha_1\right)\frac{g\theta_2}{\alpha_1\Pi_y}.$$

This term represents the development of the entrepreneur's informational advantage, conditioned on the amount of control in the next period, $g\theta_2$. Moreover, the informational agency costs depend on the difference between the ratio of beliefs α_1/α_2 and the original beliefs, α_1. We can, therefore, interpret the term in brackets as an "informational" mark-up factor.

As a next step, we investigate which of these dynamical effects has the strongest influence on S_1^{LT} under long-term contracting. As we have seen in equation (4.70), the intertemporal profit share S_1^{LT} consists of the short-term first-period

4.3 Venture capital financing and product market competition

share S_1, as well as of three additional effects.[17] We have to distinguish between two scenarios: On one hand, the first-period share under long-term contracting could be higher than (or equal to) its equivalent under short-term contracting, $S_1^{LT} \geq S_1$. In this case, the increase in agency costs and the information rent together dominate the negative competition effect of (4.70). On the other hand, the negative competition effect might have a stronger impact on S_1^{LT} than the additional agency costs and the information rent. In that case, the first-period share of the entrepreneur will actually be reduced below that under short-term contracting, $S_1^{LT} < S_1$. We, thus, derive the following corollary:

Corollary 4.1 (Intertemporal incentive problem)

1. *In case $S_1^{LT} \geq S_1$, the intertemporal agency costs and the information rent have a dominating influence on S_1^{LT}, and the firm, therefore, obtains a higher first-period incentive share of profits than under short-term contracting ("normal case").*

2. *In case $S_1^{LT} < S_1$, the firm, under long-term contracting, actually obtains a lower first-period share of profits. This surprising result is due to the strong impact of the competition effect on S_1^{LT}, which helps to realign the incentives of the entrepreneur in the intertemporal setting ("disciplinary case").*

The competition effect becomes stronger with higher potential profits from the second-period market expansion $E(\Pi_{t=2})$ and with a higher share S_{2Duo}^{LT} accruing to the entrepreneur in that stage.

In case the competition effect is reduced to zero, our results incorporate the findings of Bergemann and Hege (1998), who state that, under long-term contracting, the first-period share of the entrepreneur increases due to intertemporal agency costs and an informational rent.[18]

---

[17] The size of these three effects depends on the profit levels Π_y^A and Π_y which, in turn, are influenced by the demand parameter T, by the marginal production costs c, and by the difference of the marginal production costs between the two firms, Δc (see 4.3.1 above).

[18] If $-S_{2Duo}^{LT} \frac{E(\Pi_{t=2})}{\Pi_y} = 0$, then the first period share is equal to $S_1^{LT} = S_1 + \frac{g\theta}{\Pi_y} + (\frac{\alpha_1}{\alpha_2} - \alpha_1)\frac{g\theta}{\alpha_1 \Pi_y}$, which implies that $S_1^{LT} > S_1$, and that the scenario of Corollary 3.1.1) is in place. This result is analogous to Bergemann and Hege (1998), who show that in an intertemporal set-up, the first-period incentive share is higher than in the static framework (p.718 and Figure 2).

The intertemporal participation problem

We have stated above that, under long-term contracting, the venture capital company faces a single intertemporal participation constraint instead of the three static participation constraints. Financing for the young firm becomes easier now, since the intertemporal participation constraint allows to exchange profit shares between all three stages. The intertemporal participation constraint is given as:

$$[(1-S_1^{LT})\alpha_1\theta_1\Pi_y - g\theta_1]$$
$$+ \alpha_1\theta_1[(1-S_{2Duo}^{LT})E(\Pi_{t=2}) - g\theta_2^+] \quad (4.68')$$
$$+ (1-\alpha_1\theta_1)[(1-S_2^{LT})\alpha_2\theta_2\Pi_y - g\theta_2] \geq 0.$$

The first term of (4.68') is identical to the first-period participation constraint under short-term financing, except for the expression $(1-S_1)$, which is substituted by the long-term share $(1-S_1^{LT})$. The second and third terms represent the respective participation constraints of the second-period expansion and market entry phase, which are identical to those under short-term contracting. Yet they are weighted with their entrance probabilities $\alpha_1\theta_1$ and $1-\alpha_1\theta_1$. Since the long-term share $(1-S_1^{LT})$ can be either higher or lower than the short-term share $(1-S_1)$, we proceed to analyze the project's total value under long-term financing.

Project value under long-term contracting

In order to solve for the project value under long-term contracting, we insert the entrepreneur's minimum incentive compatible shares (4.67a'), (4.67b') and (4.70) into the respective first and second-period shares of the intertemporal participation constraint of the venture capital company (4.68) (see Appendix 4.5.4):

$$\underbrace{[\alpha_1\theta_1\Pi_y - 2g\theta_1]}_{I} - \underbrace{\left(\frac{\alpha_1}{\alpha_2} - \alpha_1\right)g\theta_1\theta_2}_{II} + $$
$$+ \alpha_1\theta_1\underbrace{[E(\Pi_{t=2}) - 2g\theta_2^+]}_{III} + (1-\alpha_1\theta_1)\underbrace{[\alpha_2\theta_2\Pi_y - 2g\theta_2]}_{IV} \geq 0. \quad (4.71)$$

Term *I* of (4.71) represents the net profits of the first period, which is identical to the minimum profit condition under short-term financing. Term *II* incorporates the informational rent (slightly modified) granted to the entrepreneur in the intertemporal context. Term *III* consists of the net profits in the expansion stage. Finally, term *IV* reflects the expected net profits of the second-period market entry stage, which is again equivalent to the second-period minimum profit condition under short-term financing. The entrepreneur obtains long-term financing for the project as long as the above expression is greater than zero.

4.3 Venture capital financing and product market competition

Next, we compare the project value under long-term contracting with its counterpart under short-term contracting. Given that short-term contracting is granted for all three scenarios, its total value is composed of:

$$[\alpha_1\theta_1\Pi_y - 2g\theta_1]$$
$$+\alpha_1\theta_1[E(\Pi_{t=2}) - S_2^{Duo}\Pi_y - 2g\theta_2^+] \qquad (4.72)$$
$$+(1-\alpha_1\theta_1)[\alpha_2\theta_2\Pi_y - 2g\theta_2].$$

Subtracting equation (4.72) from the project value under long-term contracting (4.71) yields:

$$\alpha_1\theta_1 S_2^{Duo}\Pi_y - \alpha_1(1/\alpha_2 - 1)g\theta_1\theta_2 \geq 0. \qquad (4.73)$$

Substituting $S_2^{Duo} = g/(\Pi_y^A - \Pi_y)$ leads to:

$$\alpha_1\theta_1 g\left[\frac{\Pi_y}{\Pi_y^A - \Pi_y} - \theta_2(1/\alpha_2 - 1)\right] \geq 0. \qquad (4.74)$$

The term in brackets is, generally, positive because the profit ratio is close to one, while the information term $\theta_2(1/\alpha_2 - 1)$ is rather low. Thus, for $\alpha_1, \theta_1, \theta_2^{(+)}, g \neq 0$, long-term contracting is more efficient than short-term contracting.

Now suppose for a moment that short-term financing of the project's second-period market entry stage is not profitable any more, such that refinancing is denied. Recall, however, that the project's profitability conditions under external financing are stricter than under self-financing [$\alpha_2\theta_2\Pi_y - 2g\theta_2 \geq 0$ instead of $\alpha_2\theta_2\Pi_y - g\theta_2 \geq 0$]. Thus, if the expected second-period market entry profits are drawn from the interval [$2g\theta_2 > \alpha_2\theta_2\Pi_y \geq g\theta_2$], it is socially desirable that the entrepreneur obtains financing for this stage as well. Here, a long-term contract will help to circumvent the financial restriction, if the following relation holds:

$$\alpha_1\theta_1 g\left[\frac{\Pi_y}{\Pi_y^A - \Pi_y} - \theta_2(1/\alpha_2 - 1)\right] \geq -(1-\alpha_1\theta_1)[\alpha_2\theta_2\Pi_y - 2g\theta_2], \qquad (4.75)$$

i.e. if the surplus from long-term contracting - derived in (4.74) - exceeds any potential losses from the second-period entry stage. Since we restrict our analysis to projects with a maximal loss of $[-g\theta_2]$, we can substitute $[\alpha_2\theta_2\Pi_y - 2g\theta_2]$ on the right-hand-side of (4.75) by $[-g\theta_2]$ and obtain, after rearranging terms and simplifying:

$$\frac{\Pi_y}{\Pi_y^A - \Pi_y} \geq \theta_2(1/\alpha_2 - 1) + (1 - \alpha_1\theta_1)/\alpha_1 \quad \text{for } \alpha_1, \theta_1, \theta_2, g, \neq 0. \tag{4.76}$$

Thus, if condition (4.76) is fulfilled, a long-term contract enables the entrepreneur to realize a second attempt of market entry which would not be granted under short-term financing. From a social point of view, this is desirable and the long-term contract, therefore, increases market efficiency. We summarize our results in the following proposition.

Proposition 4.14 (Long-term project financing)

(i) If the project is profitable in each stage, long-term contracting is more efficient than short-term contracting.

(ii) A long-term contract helps to circumvent financial restrictions of innovation projects that are stopped too early under short-term contracting.

Proposition 4.14 indicates that a long-term financial contract actually improves the financing situation of the young firm. The entrepreneur is better off since not only top innovation projects, but also projects of slightly lower expected value will obtain financing in both investment periods.[19] Our results also explain the widespread use of long-term contracts in equity financing relationships: Venture projects are characterized by long-term financing, in which prospective projects obtain more than one financing round (Sahlmann, 1990; Gompers, 1995; Cornelli and Yosha, 1997).

4.3.2.3 Discussion of the results

Early stopping-problem: Under short-term contracting, it can happen that a firm obtains venture capital financing for the first period of innovation, but is denied second-period financing after its project has failed. This is due to the updating of beliefs and to the lower expected value of the project in the second period. Moreover, the innovation project might, likewise, be stopped too early if *(i)* the initial beliefs about the project's quality, α_1, suddenly decrease; if *(ii)* the profit level Π_y declines due to a reduction in consumers' preferences T; or if *(iii)* the required amount of R&D investment, $g\theta_t$, increases.

The problem can be circumvented with a long-term contract: The venture capital company guarantees refinancing for the second-period entry stage in exchange for

---

[19] The result that long-term contracts may reduce inefficiencies which are caused by information problems between venture capitalists and entrepreneurs is also found in Bergemann and Hege (1998).

4.3 Venture capital financing and product market competition

potential profits from the expansion stage. [20]

Industry characteristics: Suppose that the industry into which the young firm attempts to enter is characterized by either strong consumers' preferences or distinctive product variants. This implies that the industry's parameter T is very high.[21] In that case, financing becomes easier to obtain, and both short-term and long-term financing should be available to the young firm. If, by contrast, the industry's parameter T is low, price competition between the firms will be more intense, and the expected profits from market entry will thus be lower. Financing now becomes harder to obtain. The young firm might, therefore, be forced to switch from short-term to long-term contracting.

Innovation size: As far as the size of the innovation is concerned, it is always favorable to the young firm to achieve a better cost-effectiveness than its competitor: If the innovation is drastic (Δc is large), the firm will quickly expand its market shares and increase its profit. Financing - especially for the second-period expansion stage - becomes easier to obtain.

Financial instruments: Moreover, the young firm might find a "cheaper" financing alternative than venture capital in the expansion stage. Since in that stage, the project will realize positive profits with certainty, which could serve as collateral, the firm could apply for mezzanine funds or a short-term financial institution loan (Berger and Udell, 1998). This, however, is beyond the scope of our model.[22]

---

[20] Short-term financial contracting, however, might be better for the young firm if its profits Π_y after the first-period innovation are rather high: In case the innovation project shows significant upside potential, the young firm might be able to self-finance the second period research expenditures, $g\theta_2$. This situation is given if profits are sufficiently high *(i)* to induce incentive-compatible investment, *(ii)* to compensate the venture capital company for the capital provided, and *(iii)* if sufficient financial slack remains to finance the research expenditures of the second period: i.e., if $\Pi_y - 2g/\alpha_1 \geq g\theta_2$.

[21] Sutton (1998, chapter 6) describes the market for flowmeters as an example of an industry in which high T-values are present. The different flowmeter types are characterized by their physical principles employed in measurement (electromagnetic, ultrasonic,...). A firm which discovers an alternative principle of measurement can easily enter the market and, in case of low price and production costs, will attract a large market share and realize high profits.

[22] Gompers (1998) points out that venture capital is a very costly source of funding. Thus, as soon as any tangible assets are available or a steady cash flow is realized, the young firm will switch to "cheaper" debt or mezzanine financing. In our model, however, the venture capital company realizes zero profits due to the perfect competition in the venture capital market. Its expected share of profits in the expansion stage just equals the investment costs, $g\theta_2^+$. This implies that, within our framework, there exists no "cheaper" financing alternative. Thus, we abstract from analyzing convertible securities, although recent literature has focused on these hybrid financial instruments mainly for incentive reasons (see e.g. Cornelli and Yosha, 1997).

4.3.3 Strategic reactions of the incumbent

So far we have assumed that only the young firm, but not the incumbent is able to innovate. However, the incumbent will certainly react to the potential market entry of the young firm. In this case, he can pursue different strategies. He can

- either invest in R&D activities, too, in order to reduce his marginal production costs, lower the product price and thus increase his market shares and profits.

- Or, he can invest in predatory activities in order to reduce the innovation success probability of the young firm. In that case, the incumbent has to weight up the predation costs against the additional profits from remaining a monopolist.

We now analyze this strategic behavior of the incumbent, its impact on product market competition and on the financial contract between the entrepreneur and the venture capitalist in the following two subsections.

4.3.3.1 Innovation and cost reduction

Research activities of the incumbent

The assumptions A1-A7 about the players, the time horizon, and the information structure remain basically the same as in subsection (4.3.2), except for assumption A5, which concerns the incumbent's production and innovation technology: At the beginning of the game, the incumbent produces with high marginal costs. Since we assume that the financial contract between the young firm and the venture capitalist is observable by all players, the incumbent is informed about the young firm's attempt to enter his monopoly market. Therefore, he immediately starts to invest in R&D activities too in order to reduce his marginal production costs. The R&D technology is the same as the one of the young firm. We denote the incumbent's success probability by λ_t to clearly distinguish it from the young firm's success probability θ_t. The associated research costs $g(\lambda_t)=g\lambda_t$ in each period $t=1,2$, are linearly increasing in the success probability λ_t. This success probability is restricted to the same technology $\lambda_t \in [0, \lambda^{max}]$ as the one of the young firm. In case of success, the incumbent reduces his marginal costs by Δc. Denoting the initial level of his marginal costs by c_i^h, these marginal costs will be reduced from a high to a middle and further to a low level: $c_i^h - \Delta c \rightarrow c_i^m - \Delta c \rightarrow c_i^l$, if the incumbent innovates in both competition periods.

The competition game, again, consists of two steps in every period: First, the young firm and the incumbent simultaneously choose their level of R&D

4.3 Venture capital financing and product market competition

expenditures. Nature, then, decides upon success or failure of the innovation projects. Second, both firms select their optimal prices according to their respective marginal costs, sell their products and realize their profits.

Payoffs in the two-period simultaneous move game

We begin by analyzing the gross payoffs of the young firm and the incumbent in the first period of the game: If the innovation project of the young firm is successful, the young firm enters the market with an initial cost level of c_y^h.[23] However, the young firm will only realize her level of symmetric gross profits of Π_y, if at the same time the incumbent fails to innovate. If, by contrast, the innovative activities of the incumbent are successful too, the incumbent will enjoy a competitive advantage and produce with lower marginal costs. His profits are $\Pi_i^A = \frac{1}{4}T + \frac{\Delta c}{3} + \left[\frac{\Delta c}{3}\right]^2/T$, which are higher than the entrepreneur's in case of entry.[24] Thus, our assumptions about the initial cost levels incorporate a potential first-mover advantage for the incumbent firm: They imply that if the incumbent strategically reacts to the entry of the young firm by spending on R&D himself, it will be much harder for the young firm to introduce the new product into the market. Hence, the price competition is tougher and the entrepreneur will realize a lower level of gross profits, $\Pi_y^D = \frac{1}{4}T - \frac{\Delta c}{3} + \left[\frac{\Delta c}{3}\right]^2/T$. If, on the other hand, the young firm fails to enter the market, the incumbent remains a monopolist and obtains gross profits of $\Pi_i^M(c_i^m) = \bar{s} - \frac{1}{2}T - c_i^m$ in case of successful process innovation, and $\Pi_i^M(c_i^h) = \bar{s} - \frac{1}{2}T - c_i^h$ in case his innovation fails. The first-period innovation game and the respective payoffs for the young firm and the incumbent are summarized in the following figure:

---

[23] Most likely, a venture-capital backed firm would enter the market with a cost-advantage. However, since we are interested into the marginal case in which venture capital financing is still available, we make this unfavorable assumption that the young firm initially has to produce with high marginal costs.
[24] See Appendix 4.5.1 below for the derivation of the respective profit levels.

180   4 Venture capital financing and strategic competition

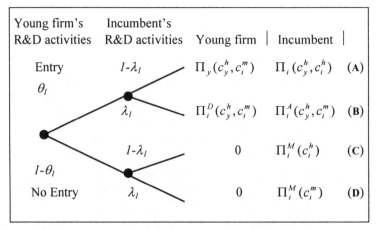

Figure 4.6: First-period payoffs when both firms spend on R&D activities

Due to the four possible outcomes of the innovation game in the first period (A)–(D), we must distinguish between four different settings in the second period.

In case (A), both the young firm and the former incumbent have realized the same marginal costs in the first period. Thus, if in the second period both firms innovate or both firms fail, they produce again with identical marginal costs and obtain symmetric gross profits, Π_y and Π_i. However, if only one of the firms successfully innovates in the second period, then only this firm reduces the marginal production costs, increases its market share and realizes a higher profit level, Π^A. The opponent, by contrast, produces with higher costs and consequently obtains a lower profit level, Π^D. These second-period payoffs following a symmetric cost-situation in the first period are given in Table 4.2:

CASE (A): $y \setminus i$	Success λ_2	Failure $1-\lambda_2$
Success θ_2	Π_y ; Π_i	Π_y^A ; Π_i^D
Failure $1-\theta_2$	Π_y^D ; Π_i^A	Π_y ; Π_i

Table 4.2: Second-period payoffs after symmetric costs in $t=1$.

In scenario (B), the innovation game of the first period has resulted in a cost-disadvantage for the young firm and a cost-advantage for the incumbent. In the second period, the young firm can at most attain a symmetric market position, if she innovates and the incumbent does not. However, if both firms either fail or innovate, the cost-disadvantage of the young firm remains, which results in a profit level of Π_y^D for the young firm, and Π_i^A for the incumbent. The worst case of scenario (B) for the young firm will be if she fails to innovate while the

4.3 Venture capital financing and product market competition

incumbent innovates again and realizes a double cost-advantage. According to our model of the circular city, the profits in this case amount to $\Pi_y^{DD} = \frac{1}{4}T - \frac{2}{3}\Delta c + [\frac{2}{3}\Delta c]^2 / T$ for the young firm (DD stands for double cost-disadvantage), whereas the incumbent obtains $\Pi_i^{AA} = \frac{1}{4}T + \frac{2}{3}\Delta c + [\frac{2}{3}\Delta c]^2 / T$ (AA stands for double cost-advantage), respectively:

CASE (B): $y \setminus i$	Success λ_2	Failure $1-\lambda_2$
Success θ_2	Π_y^D ; Π_i^A	Π_y ; Π_i
Failure $1-\theta_2$	Π_y^{DD} ; Π_i^{AA}	Π_y^D ; Π_i^A

Table 4.3: Second-period payoffs following a cost-disadvantage of the young firm in $t=1$.

In scenario (C), the young firm failed to enter in the first period. The incumbent, on the other hand, has been unable to reduce his marginal production costs in the first period. Therefore, the second-period game that both parties play has the same payoff structure as the one in the first period (cf. Figure 4.5). If the young firm fails to enter a second time, the incumbent will continue to earn monopoly profits. These monopoly profits will equal $\Pi_i^M(c_i^m) = \bar{s} - \frac{1}{2}T - c_i^m$, if the incumbent innovates, and $\Pi_i^M(c_i^h) = \bar{s} - \frac{1}{2}T - c_i^h$ otherwise:

CASE (C): $y \setminus i$	Success λ_2	Failure $1-\lambda_2$
Success θ_2	Π_y^D ; Π_i^A	Π_y ; Π_i
Failure $1-\theta_2$	0 ; $\Pi_i^M(c_i^m)$	0 ; $\Pi_i^M(c_i^h)$

Table 4.4: Second-period payoffs after the young firm failed to enter in $t=1$ and the incumbent produced with high marginal costs.

Case (D) represents the last and worst scenario for the young firm: In the first period, she has failed to enter, while the incumbent has successfully seized the opportunity to reduce his marginal costs. If the young firm now enters the market in the second period, she has to compete with a rival whose costs are lower than her own. This implies that the young firm's product can acquire only a small market share, and the respective profits in this case, Π_y^{DD} or Π_y^D, are rather modest. If, by contrast, the young firm fails again to enter the market, the incumbent will be able to realize even higher monopoly profits due to his favorable cost structure: $\Pi_i^M(c_i^l) = \bar{s} - \frac{1}{2}T - c_i^l$ in case of success, and $\Pi_i^M(c_i^m) = \bar{s} - \frac{1}{2}T - c_i^m$ in case of failure:

CASE (D): $y \setminus i$	Success λ_2	Failure $1-\lambda_2$
Success θ_2	Π_y^{DD} ; Π_i^{AA}	Π_y^D ; Π_i^A
Failure $1-\theta_2$	0 ; $\Pi_i^M(c_i^l)$	0 ; $\Pi_i^M(c_i^m)$

Table 4.5: Second-period payoffs after the young firm failed to enter in $t=1$ and the incumbent produced with medium marginal costs.

Having derived the payoffs for the young firm and the incumbent for all second-period subgames, we proceed by backward induction and solve for the optimal R&D intensity of the two firms. Since the R&D technology is linear, we know that it is optimal to either spend the maximal amount of research expenditures and to reach the upper limit of the success probability, or not to spend anything at all on research expenditures. To keep the analysis simple, we assume that even after the strategic innovation activities of the incumbent, it is still optimal for the young firm to spend the maximal amount $g\theta_t^{max}$ on R&D activities in each period $t=1,2$, (see Appendix 4.5.5 for a detailed description of the R&D investment conditions). In turn, the same holds true for the incumbent, since his expected profits in all scenarios of the game are equal to or higher than the expected profits of the young firm. We can therefore state:

Proposition 3.15 (R&D investment of the incumbent)

In each period $t=1,2$, the incumbent will invest the maximal amount of R&D expenditures, $g(\lambda_t) = g\lambda^{max}$.

Implications for the financial contracting

Given the structure of the game when both firms invest in cost-reducing R&D activities, we show how the new payoff structure affects the financial contract between the entrepreneur and the venture capitalist.

<u>Short-term financing</u>

The strategic R&D activities of the incumbent will influence the young firm's expected gross profits in each period. The profits depend on the innovation success probability of the incumbent and will, generally, decline. This implies,

4.3 Venture capital financing and product market competition

that the minimum profit conditions under short-term and long-term contracting become more restrictive. Due to the various cost-constellations between the two firms, the young firm must offer now five different financial contracts under short-term financing. The five conditions for the project to obtain financing in the first period (4.77), in the second-period market expansion stages (4.78), (4.79), and in the second-period entry stages (4.80), (4.81) are given as follows:

$$\alpha_1 \theta_1 [(1-\lambda)\Pi_y + \lambda \Pi_y^D] \geq 2g\theta_1 \tag{4.77}$$

CASE A: $$\theta_2 [(1-\lambda)(\Pi_y^A - \Pi_y) + \lambda(\Pi_y - \Pi_y^D)] \geq 2g\theta_2 \tag{4.78}$$

CASE B: $$\theta_2 [(1-\lambda)(\Pi_y - \Pi_y^D) + \lambda(\Pi_y^D - \Pi_y^{DD})] \geq 2g\theta_2 \tag{4.79}$$

CASE C: $$\alpha_2 \theta_2 [(1-\lambda)\Pi_y + \lambda \Pi_y^D] \geq 2g\theta_2 \tag{4.80}$$

CASE D: $$\alpha_2 \theta_2 [(1-\lambda)\Pi_y^D + \lambda \Pi_y^{DD}] \geq 2g\theta_2 \tag{4.81}$$

Because of the reduced expected profit levels, the above conditions are all more restrictive than those under short-term contracting, (4.58), (4.60), and (4.64), when the incumbent remains passive. In condition (4.77), the weighted average $(1-\lambda)\Pi_y + \lambda \Pi_y^D$ replaces the originally certain profit level Π_y. Thus, financing for the first period will be harder to obtain when the success probability of the incumbent is high.

For the second-period expansion stage, conditions (4.78) and (4.79) are more likely to be fulfilled the higher the difference in the respective profit levels, Π_y^A-Π_y, Π_y-Π_y^D, and Π_y^D-Π_y^{DD}. The difference between the profit levels increases with the cost-difference between the two firms. Thus, financing for the market expansion stages will become easier to obtain the more drastic the marginal cost-reduction of the young firm in the second period. Nevertheless, the expected profits from the market expansion stages are reduced.

The expected net profits in the second-period entry stage, likely being already negative if the incumbent does *not* invest in R&D activities (4.60), are further reduced if the incumbent actually *does* invest in R&D activities. Therefore, the above conditions (4.80) and (4.81) for second-period market entry are unlikely to be fulfilled.

<u>Long-term financing</u>

As in the previous section, a long-term contract might serve as a remedy to guarantee financing for the entrepreneur's second-period entry stages. However, if the reduced expected profits from the second-period expansion stage do not suffice to compensate the negative expected profits from the second-period entry stages, a long-term contract will not be offered. Thus, the incumbent's strategic R&D investment together with the information problems in long-term contracting may give rise to a negative total project value. By recalling the ranking of financial contracts in that financing was easiest to obtain for the second-period expansion stage, followed by short-term financing for the first-period market entry stage, then long-term financing, and, lastly, financing for the second-period entry stage, we can derive the following proposition:

Proposition 4.16 (Impact of the incumbent's R&D activities on the young firm's financial contract)

If the incumbent strategically invests in R&D activities, the expected value of the young firm's project will decline. In this case, venture capitalists choose to finance only later stage investments, or grant solely a short-term financial contract for the first period market entry stage.

This stands in accordance with the empirical fact, that when project risk is high, venture capitalists prefer to grant short-term financing.

Welfare aspects

As far as welfare aspects are concerned, social welfare will increase, if the incumbent successfully invests in R&D activities and the young firm is able to enter the market. In this case, competition between the two firms reduces the product prices, which leads to an increase in consumer surplus. On the other hand, if the incumbent deters the young firm's market entry by successfully engaging in R&D activities himself, and induces the venture capital company to deny refinancing for the young firm, welfare will decline: The incumbent will stay monopolist and will continue to charge his monopoly price. Welfare in the present scenario is slightly higher than in a monopoly in which the incumbent doesn't spend on R&D activities. Social welfare, of course, is much lower than under duopolistic price competition.

4.3.3.2 Predation

Theoretical aspects

The second important possibility for the incumbent to deter market entry of the young firm is to engage in predation (see also Poitevin 1989). This means that the incumbent competes with unfair means: The incumbent invests an amount of K^p to prey upon his rival. This reduces the young firm's success probability from θ_i to θ^p (see Bolton and Scharfstein 1990 for a similar approach), and gives the incumbent a higher chance of remaining a monopolist.

In this case, the gross payoffs in each subgame remain the same as under a "passive" incumbent. However, the young firm's success probability is lower now. The game-tree of the two-period innovation, entry and competition game under predation is given as follows:

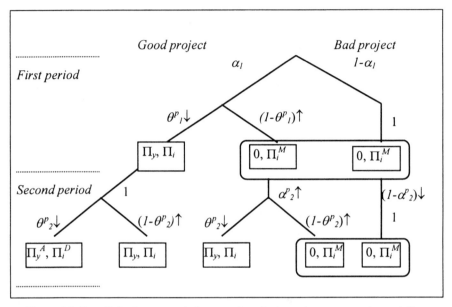

Figure 4.7: Payoffs in the two-period game when the incumbent engages in predation

In each period the incumbent chooses to prey if the additional monopoly profits are higher than the predation costs. In the first period, his calculus is given as:

$$(1-\alpha_1(\theta_1-\theta^P))\Pi_i^M + \alpha_1(\theta_1-\theta^P)\Pi_i \geq K^P$$

$$(1-\alpha_1\theta_1+\alpha_1\theta^P)\Pi_i^M + (\alpha_1\theta_1-\alpha_1\theta^P)\Pi_i \geq K^P$$

Therefore, the additional predation profits for the incumbent in the first period are equal to:

$$\alpha_1(\theta_1-\theta^P)(\Pi_i^M - \Pi_i) \geq K^P \qquad (4.82)$$

Thus, the incumbent is more inclined to engage in predation the higher the difference in profit levels between monopoly and duopoly, the higher the reduction in the young firm's success probability, $\theta-\theta^P$, the higher the initial beliefs about the young firm's project, α_1, and the lower the predation costs K^P.

In the second period, the incumbent decides to engage in predatory activities again, if the additional profits exceed his predation costs K^p. In the second entry subgame, his decision depends on

$$\alpha_2^p (\theta_2 - \theta_2^p)(\Pi_i^M - \Pi_i) \geq K^p, \tag{4.83}$$

while in the second-period expansion stage, the incumbent compares

$$(\theta_2 - \theta_2^p)(\Pi_i - \Pi_i^D) \geq K^p. \tag{4.84}$$

Note that the predation activities also affect the entrepreneur's and venture capitalist's updating of beliefs about the project's quality. The posterior beliefs under predation are derived as (see also Appendix 4.5.2)

$$\frac{\alpha_1}{\alpha_2^p} = \alpha_1 + \frac{1-\alpha_1}{1-\theta_1^p} \tag{4.85}$$

Thus, if the incumbent's predation reduces the young firm's success probability, the updated beliefs of developing a good, yet unsuccessful project are higher than under competition without predation, $\alpha_2^p > \alpha_2$.

Short-term financing

The incumbent's predation activities have the following effects on the financial contracting between the young firm and the venture capitalist: Under short-term contracting, it becomes more difficult in each scenario to obtain project financing, because research expenditures remain the same and expected profits decrease. In the first period we have

$$\alpha_1 \underset{(-)}{\theta_1^p} \Pi_y - 2g\theta_1 \geq 0. \tag{4.86}$$

In the second-period entry stage, the condition becomes more restrictive as well

$$\underset{(+)}{\alpha_2^p} \underset{(-)}{\theta_2^p} \Pi_y - 2g\theta_2 \geq 0. \tag{4.87}$$

The same holds true for the second-period expansion stage

$$\underset{(-)}{\theta_2^p} (\Pi_y^A - \Pi_y) + (1 - S_{2Duo}^p)\Pi_y - 2g\theta_2 \geq 0. \tag{4.88}$$

The ranking of the financial contracts, nevertheless, remains the same as under Proposition 4.12.

Long-term financing

Under long-term contracting and predation, the profit shares S_1^P, S_2^P, S_{2Duo}^P that the entrepreneur requires for truthfully allocating the funds, all increase:

$$S_1^P = \underbrace{\frac{g\theta_1}{\alpha_1 \theta_1^P \Pi_y}}_{\substack{I \\ (+)}} + \underbrace{\frac{g\theta_2}{\Pi_y}}_{II} - \underbrace{S_{2Duo}^P}_{(+)} \underbrace{\frac{E(\Pi_{t=2}^P)}{\Pi_y}}_{III} + \underbrace{\frac{(1-\alpha_1)}{\alpha_1 \Pi_y} \frac{g\theta_2}{(1-\theta_1^P)}}_{\substack{IV \\ (-)}}, \qquad (4.89)$$

$$\underset{(+)}{S_2^P} = g\theta_2 / \underset{(+)}{\alpha_2^P} \underset{(-)}{\theta_2^P} \Pi_y, \qquad (4.90)$$

$$\underset{(+)}{S_{2Duo}^P} = g\theta_2 / \underset{(-)}{\theta_2^P} (\Pi_y^A - \Pi_y). \qquad (4.91)$$

We see that the first term (*I*) of S_1^P, i.e. the static agency costs, increases under predation. This is intuitively clear, because the reduced innovation success probability makes it more attractive for the entrepreneur to misuse the funds instead of employing them towards project realization. The third term (*III*), our "competition effect", is increasing in S_{2Duo}^P under predation (see (4.91)). Term (*IV*) finally incorporates the information rent, which will be reduced. The information rent under predation is lower, because the updating of beliefs is accomplished more slowly, since the entrepreneur controls only part of the success probability via the investment of funds.

As far as the second-period shares are concerned that the entrepreneur requires for truthful investment, they both will raise due to the decline in expected profits under predation.

Moreover, the reduced success probability θ^P makes it harder to fulfill the participation constraint for the venture capital company. Altogether, the project's total value under long-term contracting and the incumbent's predation definitely will decline, making it more difficult for the young firm to obtain venture capital financing and to enter the market.

Thus, as in the preceding subsection, it can happen that a long-term contract is not offered any longer because the total expected project value has become negative. In that case, the best the young firm can expect is short-term provision of capital in the first period.

Welfare aspects

As far as social welfare is concerned, welfare will decrease due to the reduced entry probabilities, which imply a higher persistence of monopoly. In addition to this, welfare is reduced by the amount of predation costs, K^P, which represent a pure deadweight loss.

Empirical findings

A recent example of some incumbent's predatory activities against a young, venture capital financed firm is found in the Swiss concrete repair industry (Sidler, 1997, 191pp.): In Switzerland - as in many other countries - highway bridges are built with concrete into which metal layers are inserted for stability reasons. These metal particles, however, are subject to erosion due to rainwater filtering through, such that repair becomes necessary every 10 to 15 years. The standard procedure consists of removing the concrete bottom layer of the bridge with chisels, thereby milling out the rusty parts. After that, a new layer of concrete is sprayed onto the bridge from below. This process is tedious for the workers who get covered with dust and particles, as well as time-consuming and costly because of the putting up and down of the scaffolds.

In the beginning of the 1990s, a group of engineers from the Technical University of Zurich developed an alternative way for this highway bridge concrete repair: In order to replace the traditional method, they invented an electrochemical procedure by which an anti-erosion liquid was filtered through the concrete from above. This procedure was easier to handle (the liquid can be simply applied from above), faster, and less costly than the standard repair process. Thus, the research group had no problem to obtain venture capital financing for their project. Moreover, since the Swiss government planned to have a large part of the highway bridges and tunnels restructured by the mid 1990s, the project's prospects from the demand side were rather good.

However, plans had been made without taking strategic reactions of competitors into account. As soon as the existing firms learned about the new method developed by the young firm, they formed a coalition in order to prevent the young firm from establishing itself in the market. More precisely, incumbents began to lobby among the Swiss governments, hoping to prevent that any governmental repair business would be given to the young firm. Because of the long-time relationship between the government and the incumbent firms, as well as the missing track record of the young firm, the predation activities were successful.

The young firm was denied any major concrete repair order, and, thus, could not establish itself in the market. In Switzerland today, highway bridge concrete repair is still done according to the old, costly method, whereas the superior procedure failed to make its point, and the young firm, finally, had to withdraw from the market.

4.3.4 Discussion

In section 4.3 we presented a dynamic agency model of venture capital financing, where an entrepreneur attempted to enter a market with an innovative product. We explicitly formalized the market entry of the young firm and its subsequent price competition with an incumbent. In contrast to the existing literature, we thus endogenized the stochastic returns from the venture project. We showed that the venture project's profits were higher in industries in which consumers have strong preferences for the product variants supplied. Moreover, the venture project's profits increased when the size of innovation is drastic, i.e. when the entrepreneur could generate a substantial cost-advantage in the second-period market expansion stage.

The main results of our analysis on the interaction between venture capital financing and dynamic product market competition are as follows.

- Venture capital companies are reluctant to finance pure start-up projects and prefer to engage in expansion stage financing.
- The entrepreneur, on the other hand, seeks to obtain venture capital for both the early-stage and the expansion-stage periods, and, therefore, prefers a long-term financial contract.
- The riskier the innovation project and the fiercer the competition in the product market, the more likely, however, that venture capital is provided only via short-term contracts.

The hierarchy of financial contracts derived in our model stands in accordance with the empirical evidence that the majority of venture capital is invested into expansion stage projects and not into seed and start-up financing, which are associated with much higher risk.

Under short-term contracting, bad quality projects are presumably stopped after the first period. On the other hand, it is quite likely that good, yet unsuccessful projects are denied follow-up financing in the second period. This is due to the imperfect information about the project's quality and to the moral hazard problem between the firm and the venture capital company. It implies that a fraction of good innovation projects will be stopped prematurely, which, in turn, increases the incumbent's chances of remaining a monopolist. Our model, thus, indicates that financial market imperfections will spill over to the product market, where they will reinforce concentration tendencies within the industry.

The early-stopping problem can be circumvented via a long-term financial contract, which allows for an intertemporal trade-off of profit shares. In this case, refinancing is guaranteed for all second-period R&D investments. The long-term relationship, however, imposes additional agency costs on the financial contract.

Surprisingly, though, competition in the product market actually helps to realign the incentives of the entrepreneur: If this "competition effect" is strong enough, the moral hazard problem under long-term contracting will be reduced. This is a new result in the financing literature. We, therefore, strongly suggest that future research on venture capital contracting should take the competitive environment of young, innovative firms into account. Otherwise, the incentive problems in dynamic financial contracting may appear too strong.

As long as the innovation project is profitable in all scenarios, long-term contracts are more efficient than short-term contracts. However, asymmetric information in the venture capital relationship as well as uncertainty about the project's quality impose restrictions on the financing of the innovation project. As a consequence, venture capital financing is always less efficient than self-financing of the project.

In subsection 4.3.3 we finally assumed that the incumbent didn't merely observe the young firm's potential market entry, but that he strategically reacted to the new competitor. The incumbent's strategic behavior affects the financial contracting between the young firm and the venture capital company, and, in turn, influences the market structure of the industry.

If the incumbent engages in cost-reducing innovation, too, the outcome in the product market is characterized by various cost and price constellations. The young firm has to offer five different contracts under short-term financing. All financing conditions are stricter now due to the decline in expected profit levels. Long-term financing also becomes more difficult to obtain.

In case the young firm successfully enters and both firms innovate, product prices will be lower and social welfare will increase. If, on the other hand, the young firm cannot enter because expected profits are too low to obtain venture capital financing, the incumbent will remain monopolist, and social welfare will decrease.

If the incumbent engages in entry-deterring predation instead, the young firm's innovation success probability and its profits will decline. The ranking of the financial contracts stays the same, but all financing conditions will be more restrictive. Under these circumstances, a long-term venture capital contract may become unavailable, such that short-term venture capital may be the only source of funding for the young firm.

Since the persistence of monopoly increases when the incumbent engages in predatory activities, social welfare will decline. Moreover, predation costs represent a pure dead-weight loss for the economy.

The results of our model clearly demonstrate that a young, innovative firm doesn't live in a competitive vacuum. Instead, the success probability of the innovation project and its expected returns are influenced by the strategic reactions of competitors. This, in turn, affects the dynamic financial contracting between the entrepreneur and the venture capitalist, making it more difficult to obtain venture capital financing at all. Our theoretical findings are largely supported by empirical evidence.

The model could be extended in at least two ways:

First of all, we have not modeled the venture capitalist's selection process. The initial belief about the project's quality α_1 is exogenous. However, it would be more elegant to have it endogenously derived from the venture capitalist's due diligence process.

Moreover, our long-term venture capital contract doesn't provide any sorting mechanism which would stop bad quality projects at an early stage. We think, however, that this is of minor importance for our analysis. Venture capital companies typically prefer to refinance unsuccessful projects, and not to abandon them right away, because this strategy would result in a sure loss.

Lastly, venture capital financing is provided in the form of equity. We could have included convertible securities instead, for example in the market expansion stage, or if we had introduced a positive liquidation value after each project stage. The main aim of our model, however, was to analyze the interaction between venture capital financing and product market competition. Therefore, we did not further investigate the design of mixed-form securities as it is done in the bilateral papers on venture capital contracting.

4.4 Conclusion

In the present chapter we analyzed venture capital financing of a young firm's innovation project. Financing is provided in the form of equity capital which implies that the investor, i.e. the venture capital company, fully participates in the gains and losses of the - risky - project.

We first investigated the bilateral relationship between the start-up entrepreneur and the venture capitalist. The control-theoretic literature shows that the entrepreneur must demonstrate above-average management skills in order to retain control of the project. Otherwise, the venture capital company will search for a professional management to replace the entrepreneur. The other part of literature on bilateral venture capital contracts deals with the allocation of property rights. Here, we analyzed the impact of asymmetric information between the entrepreneur and the venture capitalist on the financial contracting.

On the industry level, we presented our own work which investigates the interaction between venture capital financing and product market competition. We here endogenized the stochastic returns of the venture project by introducing two periods of price competition during which the firms compete for market shares. Moreover, we assumed that the young firm's innovation project combines product and process innovation. We analyzed the impact of dynamic moral hazard and uncertain project quality on the financial contract. We showed that product market competition actually helps to realign the incentives of the entrepreneur towards a truthful investment of funds.

Finally, we investigated how strategic reactions of the incumbent, including predation, affected the provision of venture capital. We showed that, due to the reduction of profit expectations, venture capital contracts became much more difficult to obtain. This, in turn, reduced innovation activities and at the same time increased the concentration tendency in the industry.

4.5 Appendix

Nothing gives such weight and dignity to a book as an appendix.
Herodot (490-430 B.C.)

Appendix 4.5.1 Price competition, market shares and profits

According to the Salop-model, the profit function of firm i equals:

$$\Pi_i(p_i, p_y, c_i, T) = (p_i - c_i)\left(\frac{p_y - p_i + T/2}{T}\right). \tag{A11}$$

The first-order condition and the price reaction function are given by:

$$\frac{\partial \Pi_i(p_i, p_y, c_i, T)}{\partial p_i} = \frac{p_y + T/2 - 2p_i}{T} + \frac{c_i}{T} \stackrel{!}{=} 0, \text{ and}$$

$$p_i = \tfrac{1}{2} \cdot (p_y + T/2 + c_i). \tag{A12a}$$

The price reaction function of the young firm y is analogously given by

$$p_y = \tfrac{1}{2} \cdot (p_i + T/2 + c_y). \tag{A12b}$$

The optimal prices are found at the intersection of the reaction functions:

$$p_i^*(c_i, c_y, T) = \tfrac{1}{2}T + \tfrac{1}{3}(2c_i + c_y), \text{ and} \tag{A13a}$$

$$p_y^*(c_i, c_y, T) = \tfrac{1}{2}T + \tfrac{1}{3}(2c_y + c_i). \tag{A13b}$$

4.5 Appendix

Resubstitution into the firm-specific demand function gives us the respective market shares:

$$D_i(c_i,c_y,T) = \frac{p_y + T/2 - p_i}{T} = \frac{\frac{1}{2}T + \frac{1}{3}c_y - \frac{1}{3}c_i}{T} = \frac{1}{2} + \frac{1}{3T}(c_y - c_i), \quad \text{(A14a)}$$

$$D_y(c_i,c_y,T) = 1 - D_i(c_i,c_y,T). \quad \text{(A14b)}$$

Finally, the gross profits in reduced form are derived as:

$$\Pi_i(c_i,c_y,T) = (p_i - c_i)D_i(c_i,c_y,T). \quad \text{(A15)}$$

If both firms produce with symmetric costs $c_y = c_i$, optimal product prices are given by

$$p^*(c_i,c_y,T) = \tfrac{1}{2}T + c; \quad \text{(A16)}$$

market shares equal one half,

$$D_i(c_i,c_y,T) = D_y(c_i,c_y,T) = 1/2; \quad \text{(A17)}$$

and profits are given by,

$$\Pi_i(c_i,c_y,T) = (\tfrac{1}{2}T + \tfrac{1}{3}(2c_i + c_y) - c_i) \cdot \tfrac{1}{2} = \Pi_y(c_i,c_y,T) = \tfrac{1}{4}T. \quad \text{(A18)}$$

Therefore, the profits in the symmetric case depend only on the size of the consumers' preferences, T.

If the two firms produce with different marginal costs, we denote this cost difference by Δc. Then, the profits of the firm who produces with lower marginal costs (= firm y) are equal to:

$$\begin{aligned}\Pi_y^A(c_i,c_y,T) &= (p_y - c_y)D_y = [\tfrac{1}{2}T - \tfrac{1}{3}(-\Delta c)][\tfrac{1}{2} + (\Delta c/3)/T] \\ &= T/4 + \Delta c/3 + (\Delta c/3)^2/T.\end{aligned} \quad \text{(A19)}$$

Thus, the profits for the firm with the cost-advantage are positively influenced by Δc. The profits for the firm who has a cost-disadvantage (presumably firm i) are equal to:

$$\begin{aligned}\Pi_i^D(c_i,c_y,T) &= (p_i - c_i)D_i = [\tfrac{1}{2}T - \tfrac{1}{3}(\Delta c)][\tfrac{1}{2} - (\Delta c/3)/T] \\ &= T/4 - \Delta c/3 + (\Delta c/3)^2/T\end{aligned} \quad \text{(A20)}$$

This profit level is negatively influenced by the cost-difference between the two firms.

Appendix 4.5.2 Learning process

The updating of the beliefs is accomplished according to the Bayes-rule:

$$\alpha_2 = \frac{\alpha_1(1-\theta_1)}{\alpha_1(1-\theta_1)+1-\alpha_1} = \frac{\alpha_1(1-\theta_1)}{1-\alpha_1\theta_1}. \qquad (A21)$$

Rearranging terms shows that

$$1-\alpha_1\theta_1 = \frac{\alpha_1}{\alpha_2}(1-\theta_1). \qquad (A22)$$

Solving for (α_1/α_2) and taking the inverse ratio gives us:

$$\frac{\alpha_1}{\alpha_2} = \alpha_1 + (1-\alpha_1)\frac{1}{(1-\theta_1)}. \qquad (A23)$$

(A23) again shows that the initial beliefs, α_1, are larger than the updated beliefs, α_2.

Appendix 4.5.3 Minimum share for truthful investment

To prevent the entrepreneur from moral hazard under long-term contracting, the share of profits under truthful investment must be at least equal to the payoffs under deviation. Inserting the full expressions into (4.67c'), we obtain equation (4.69):

$$\alpha_1\theta_1 S_1^{LT}\Pi_y + \alpha_1\theta_1 S_{2Duo}^{LT} E(\Pi_{t=2}) + (1-\alpha_1\theta_1)(\alpha_2\theta_2 S_2^{LT}\Pi_y)$$
$$\overset{!}{=} g\theta_1 + \frac{\alpha_1}{\alpha_2}(\alpha_2\theta_2 S_2^{LT}\Pi_y). \qquad (4.69)$$

Rearranging terms leads to:

$$\alpha_1\theta_1 S_1^{LT}\Pi_y = g\theta_1 - \alpha_1\theta_1 S_{2Duo}^{LT} E(\Pi_{t=2}) + (\alpha_2\theta_2 S_2^{LT}\Pi_y)\left[\frac{\alpha_1}{\alpha_2} - (1-\alpha_1\theta_1)\right]$$

$$= g\theta_1 - \alpha_1\theta_1 S_{2Duo}^{LT} E(\Pi_{t=2}) + (\alpha_2\theta_2 S_2^{LT}\Pi_y)\frac{\alpha_1}{\alpha_2}\left[1 - \frac{(1-\alpha_1\theta_1)\alpha_2}{\alpha_1}\right]. \quad (A24)$$

Substituting α_2 by (A21) for the term in the brackets yields:

$$\alpha_1\theta_1 S_1^{LT}\Pi_y = g\theta_1 - \alpha_1\theta_1 S_{2Duo}^{LT} E(\Pi_{t=2}) + (\alpha_2\theta_2 S_2^{LT}\Pi_y)\frac{\alpha_1}{\alpha_2}\theta_1. \qquad (A25)$$

Moreover, we multiply the entrepreneur's value under deviation from (4.67c)

$$V^D(\alpha_1) = g\theta_1 + \frac{\alpha_1}{\alpha_2}(\alpha_2\theta_2 S_2^{LT}\Pi_y), \qquad (4.67c)$$

by θ_1, and rearrange terms in order to obtain the following expression:

$$\frac{\alpha_1}{\alpha_2}\theta_1(\alpha_2\theta_2 S_2^{LT}\Pi_y) = \theta_1 V^d(\alpha_1) - g\theta_1^2. \qquad (A26)$$

Inserting this expression into the above equation (A25), we get:

$$\begin{aligned}\alpha_1\theta_1 S_1^{LT}\Pi_y &= g\theta_1 - \alpha_1\theta_1 S_{2Duo}^{LT} E(\Pi_{t=2}) + \theta_1 V^d(\alpha_1) - g\theta_1^2 \\ &= (1-\theta_1)g\theta_1 - \alpha_1\theta_1 S_{2Duo}^{LT} E(\Pi_{t=2}) + \theta_1 V^d(\alpha_1).\end{aligned} \qquad (A27)$$

Next, we take the modified profits under deviation, $V^{d\prime} = g\theta_1 + g\theta_2(\alpha_1/\alpha_2)$ from (4.67c'). We substitute for (α_1/α_2) by (A23) and obtain:

$$V^d(\alpha_1) = g\theta_1 + g\theta_2\left[\alpha_1 + \frac{(1-\alpha_1)}{(1-\theta_1)}\right]. \qquad (A28)$$

Inserting this expression for $V^d(\alpha_1)$ into equation (A27), we derive:

$$\begin{aligned}\alpha_1\theta_1 S_1^{LT}\Pi_y &= (1-\theta_1)g\theta_1 - \alpha_1\theta_1 S_{2Duo}^{LT} E(\Pi_{t=2}) \\ &\quad + \theta_1\left[g\theta_1 + g\theta_2\alpha_1 + g\theta_2\frac{(1-\alpha_1)}{(1-\theta_1)}\right].\end{aligned} \qquad (A29)$$

After rearranging and collecting terms, we obtain:

$$\alpha_1\theta_1 S_1^{LT}\Pi_y = g\theta_1 + \alpha_1\theta_1 g\theta_2 - \alpha_1\theta_1 S_{2Duo}^{LT} E(\Pi_{t=2}) + \theta_1 g\theta_2 \frac{(1-\alpha_1)}{(1-\theta_1)}. \qquad (A30)$$

We divide the above equation by $\alpha_1\theta_1\Pi_y$ to solve for the minimum share S_1^{LT} of profits that the entrepreneur requires for truthful investment in the first period of the long-term contract:

$$S_1^{LT} = \frac{g}{\alpha_1\Pi_y} + \frac{g\theta_2}{\Pi_y} - S_{2Duo}^{LT}\frac{E(\Pi_{t=2})}{\Pi_y} + \frac{(1-\alpha_1)}{\alpha_1\Pi_y} \cdot \frac{g\theta_2}{(1-\theta_1)}. \qquad (A31)$$

Finally, we substitute (A23) into the last term of the equation above and obtain:

$$S_1^{LT} = \frac{g}{\alpha_1 \Pi_y} + \frac{g\theta_2}{\Pi_y} - S_{2Duo}^{LT} \frac{E(\Pi_{t=2})}{\Pi_y} + \left(\frac{\alpha_1}{\alpha_2} - \alpha_1\right) \frac{g\theta_2}{\alpha_1 \Pi_y}. \tag{A32}$$

Q.E.D.

Appendix 4.5.4 Project value under long-term contracting

To calculate the expected minimum project value that is required under long-term contracting, we insert the incentive compatible shares of the entrepreneur that we derived in Proposition 4.13 into the intertemporal participation constraint of the venture capital company:

$$[(1 - S_1^{LT})\alpha_1 \theta_1 \Pi_y - g\theta_1] + \alpha_1 \theta_1 [(1 - S_{2Duo}^{LT})E(\Pi_{t=2}) - g\theta_2^+]$$
$$+ (1 - \alpha_1 \theta_1)[(1 - S_2^{LT})\alpha_2 \theta_2 \Pi_y - g\theta_2] \geq 0. \tag{4.58'}$$

As a first step, we insert the second period minimum share $S_2^{LT} = g/\alpha_2 \Pi_y$ from (4.67b') into the last term of (4.58') and obtain after simplifying:

$$[(1 - S_1^{LT})\alpha_1 \theta_1 \Pi_y - g\theta_1] + \alpha_1 \theta_1 [(1 - S_{2Duo}^{LT})E(\Pi_{t=2}) - g\theta_2^+]$$
$$+ (1 - \alpha_1 \theta_1)[\alpha_2 \theta_2 \Pi_y - 2g\theta_2] \geq 0. \tag{A33}$$

Next, we substitute $S_{2Duo}^{LT} = g/(\Pi_y^A - \Pi_y)$ from (4.67a') in the second term of (A33) and rearrange to get:

$$[(1 - S_1^{LT})\alpha_1 \theta_1 \Pi_y - g\theta_1] + \alpha_1 \theta_1 \left[E(\Pi_{t=2}) - 2g\theta_2^+ - g\frac{\Pi_y}{\Pi_y^A - \Pi_y}\right]$$
$$+ (1 - \alpha_1 \theta_1)[\alpha_2 \theta_2 \Pi_y - 2g\theta_2] \geq 0. \tag{A34}$$

Finally, we have to insert the first period minimum share from (4.70)

$$S_1^{LT} = \frac{g}{\alpha_1 \Pi_y} + \frac{g\theta_2}{\Pi_y} - S_{2Duo}^{LT} \frac{E(\Pi_{t=2})}{\Pi_y} + \left(\frac{\alpha_1}{\alpha_2} - \alpha_1\right) \frac{g\theta_2}{\alpha_1 \Pi_y} \tag{4.70}$$

into the first term of (A34). Before doing this, we substitute $S_{2Duo}^{LT} = g/(\Pi_y^A - \Pi_y)$ in (4.70) as well and write $E(\Pi_{t=2}) = \theta_2^+ (\Pi_y^A - \Pi_y) + \Pi_y$ in full length in order to get (4.70')

$$S_1^{LT} = \frac{g}{\alpha_1 \Pi_y} + \frac{g\theta_2}{\Pi_y} - \frac{g}{(\Pi_y^A - \Pi_y)} \frac{[\theta_2^+ (\Pi_y^A - \Pi_y) + \Pi_y]}{\Pi_y} + \left(\frac{\alpha_1}{\alpha_2} - \alpha_1\right) \frac{g\theta_2}{\alpha_1 \Pi_y}.$$

(4.70')

Inserting this into (A34) we obtain:

$$\alpha_1\theta_1\Pi_y - g\theta_1 - g\theta_1 - g\theta_2\alpha_1\theta_1$$
$$+ \frac{g}{(\Pi_y^A - \Pi_y)}[\theta_2^+(\Pi_y^A - \Pi_y) + \Pi_y]\alpha_1\theta_1 - \left(\frac{\alpha_1}{\alpha_2} - \alpha_1\right)g\theta_2\theta_1$$
$$+ \alpha_1\theta_1\left[E(\Pi_{t=2}) - 2g\theta_2^+ - g\frac{\Pi_y}{\Pi_y^A - \Pi_y}\right] + (1-\alpha_1\theta_1)[\alpha_2\theta_2\Pi_y - 2g\theta_2] \geq 0. \quad \text{(A35)}$$

If $\theta_1 = \theta_2^{(+)} = \theta^{\max}$, simplifying yields straightforwardly equation (4.71):

$$[\alpha_1\theta_1\Pi_y - 2g\theta_1] - \left(\frac{\alpha_1}{\alpha_2} - \alpha_1\right)g\theta_1\theta_2$$
$$+ \alpha_1\theta_1[E(\Pi_{t=2}) - 2g\theta_2^+] + (1-\alpha_1\theta_1)[\alpha_2\theta_2\Pi_y - 2g\theta_2] \geq 0. \quad (4.71)$$

Q.E.D.

Appendix 4.5.5 *R&D investment conditions for the young firm*

If the incumbent strategically reacts to the entry of the young firm and equivalently invests in R&D activities, the expected profits of the young firm will be dependent on the success probability of the incumbent and will decline. We therefore have to control for each marginal R&D investment condition, i.e. we have to check for each scenario whether the additional profits from innovation are higher than the marginal research costs.

The R&D investment condition that has to be fulfilled in the first period is given by:

$$\alpha_1\left[(1-\lambda_1)\Pi_y + \lambda_1\Pi^D\right] \geq g \quad \text{(A36)}$$
$$\Leftrightarrow \alpha_1\{[1-\lambda_1]\tfrac{1}{4}T + \lambda_1[\tfrac{1}{4}T - \Delta c/3 + (\Delta c/3)^2/T]\} \geq g$$

On the right hand side of (A36) we have substituted the profits by their expressions in terms of transportation costs and the cost differences levels, that we derived earlier in our the model of price competition with horizontally differentiated products.

In the second period, we have to distinguish between the four possible scenarios. The R&D investment condition in case (A), i.e. if both firms have produced with identical marginal costs in the first period, is given as:

$$(1-\lambda_2)(\Pi^A - \Pi_y) + \lambda_2(\Pi_y - \Pi^D) \geq g \tag{A37}$$

$$\Leftrightarrow (1-\lambda_2)[\Delta c/3 + (\Delta c/3)^2 / T] + \lambda_2[\Delta c/3 - (\Delta c/3)^2 / T] \geq g$$

This condition is the easier fulfilled the higher the probability of failure for the incumbent, *(1-λ₂)*, the higher the transportation costs (i.e. the stronger the consumers' preferences), *T*, the higher the potential cost-difference, *Δc*, and the lower the marginal research costs, *g*.

In case (B), the young firm has entered and produced with a cost-disadvantage in the first period. The investment condition for the second period is therefore given as:

$$(1-\lambda_2)(\Pi_y - \Pi^D) + \lambda_2(\Pi^D - \Pi^{DD}) \geq g \tag{A38}$$

$$\Leftrightarrow (1-\lambda_2)[\Delta c/3 - (\Delta c/3)^2 / T] + \lambda_2[\Delta c/3 - (\Delta c/3)^2 \cdot 3/T] \geq g$$

Again, the condition depends positively on the transportation costs, *T*, and negatively on the marginal research costs, *g*. However, this condition is harder to fulfill. Potential profits from innovation for the young firm are rather modest, because the incumbent has already achieved a cost-advantage in the first period. Thus, it might be that the young firm decides not to spend anything on R&D, because marginal research costs are higher than marginal profits. In this special second period scenario, the young firm, although she operates in the market, refrains from innovation activities, saves on R&D expenditures, but produces with higher marginal costs than the incumbent. Therefore, she supplies to a small market share only and realizes low expected profits.

In case (C), the young firm has not entered the market in the previous period. The R&D investment condition for second-period entry equals:

$$\alpha_2[(1-\lambda_2)\Pi_y + \lambda_2\Pi^D] \geq g \tag{A39}$$

$$\Leftrightarrow \alpha_2\{\tfrac{1}{4}T - \lambda_2[\Delta c/3 - (\Delta c/3)^2 / T]\} \geq g$$

This R&D investment condition is the easier fulfilled the higher the a-posteriori beliefs about the innovation project's quality, α_2, and the stronger the consumers' preferences for the product variants, *T*. Moreover, condition (A39) holds the easier, the higher the probability of failure for the incumbent, *1-λ₂*, the lower the marginal research costs, *g*, and the *lower* the cost-difference between the two firms, *Δc* (since it is the incumbent who realizes the cost-advantage)!

In case (D), the young firm has not entered the market in the first period, but the incumbent has successfully reduced his marginal production costs. The R&D investment condition thus is:

$$\alpha_2[(1-\lambda_2)\Pi^D + \lambda_2\Pi^{DD}] \geq g. \tag{A40}$$

$$\Leftrightarrow \alpha_2\{\tfrac{1}{4}T - (\Delta c/3)(1+\lambda_2) + (\Delta c/3)^2(1+3\lambda_2)/T]\} \geq g.$$

This last condition is stricter than condition (A39). However, as the one above, it depends positively on the beliefs α_2, and on the transportation costs, T, whereas it depends negatively on the success probability of the opponent, λ_2, and on the marginal research costs, g.

5 Conclusion

In the present analysis we assume that financial markets are imperfect due to asymmetric information. The information problem imposes agency costs on the financial contracting between a firm and its investor. The imperfections in financial markets will spill over to the product market, where they affect the firms' innovation activities as well as their pricing strategies.

If firms need external financing, concentration in the output market tends to rise. The reverse causality is also given: If profit opportunities in the output market change, the terms of the financial contract will also be altered. We, therefore, can conclude that financial market and product market imperfections reinforce each other.

We further illustrate this result by comparing the debt financing of chapter 3 to the equity financing of chapter 4. We investigate how these financial contracts in the input market will affect strategic competition in the output market.

On the product market, we use the same models from industrial organization to characterize the firms' competitive strategies: Both firms produce heterogeneous products and compete in prices for market shares. The time horizon is two periods of subsequent competition in the product market.

In order to acquire a certain market position, firms additionally engage in R&D activities. In case of debt financing, firms attempt to realize process innovations in each period in order to cut their marginal production costs. In case of equity financing, the aim of these R&D activities is to realize a product innovation, combined with a process innovation later on.

The competition models differ in the initial market structure. In chapter 3, we consider a duopoly of two well-established firms, in which one or both firms need external debt financing. The conditions of the debt contract, however, may force one firm to leave the market. Thus, the market structure here may change from duopoly to monopoly. In chapter 4, by contrast, we assume that the venture capital financed firm wants to enter a self-financed incumbent's product market. That is, market structure switches from monopoly to duopoly.

We now compare the features of both types of contracts more in detail before we analyze their respective impact on product market competition and innovation.

Financial contracting in the input market

We compare the debt contract from chapter 3 to the equity contract from chapter 4 with respect to the following points: *(i)* Contracting parties; *(ii)* information problem; *(iii)* contract design; and *(iv)* agency costs of the financial contracting.

(i)   Contracting parties

We analyze the financial contracts in the framework of a principal-agent relationship. On the debt-financing side, two independent banks provide credit for two well-established firms. The banking sector is characterized by perfect competition. Here, we assume that banks have all the bargaining power. This assumption merely serves to obtain clear results; bargaining power in financial contracting could as well be on the side of the firms. The firms, however, have some internal funds which they need for financing their R&D activities. The debt contract is necessary to the finance fixed production costs, e.g. to finance machinery or buildings.

On the equity side, a venture capital company as the principal provides funds to finance an entrepreneur's innovation project. The venture capital industry is also characterized by perfect competition. The entrepreneur as the agent owns the innovative project idea and is completely wealth-constrained. Nevertheless, all bargaining power is on the side of the entrepreneur, because good project ideas are rare and equity capital is abundant.

(ii)   Information problem

The information problem in the credit market is only one-dimensional. Here, we assume that the bank cannot observe the entrepreneur's investment of funds. Thus, if the innovation fails, the bank is unable to tell whether this is due to bad luck, i.e. nature, or whether the entrepreneur has diverted the funds to her private ends. Banks, too, have no monitoring technology available to encounter the entrepreneur's moral hazard problem.

The information problem in venture capital contracting is two-fold. First of all, the innovation project's quality is uncertain. Yet all players share a common initial belief that the project is of good quality. More information arrives by developing the project, and quality estimations are updated according to the Bayes-rule (learning process). Second, the entrepreneur is induced to dynamic moral hazard, because the venture capital company cannot observe the allocation of funds. If the entrepreneur doesn't invest the funds in the discovery process, the project's success probability will be zero. In this case, the learning process about the project's quality will also become asymmetric. The venture capital company also has no monitoring technology to reduce the informational asymmetries.

(iii) Contract design

The design of the financial contracts is rather complex. Therefore, we individually compare for each type of contract the features *(a)* amount of financing needs; *(b)* return distribution; and *(c)* refinancing probabilities.

(a) Amount of financing needs

In case of debt financing, firms need to borrow the amount of fixed production costs in each period. Financing is provided via long-term debt contracts. The essential features of these long-term debt contracts are fixed repayments and refinancing probabilities which are based on reported profits.

In case of equity financing, venture capital is needed to finance the R&D expenditures in both periods. Under short-term financing, there exists a hierarchy of contracts which states that financing is easiest to obtain for the expansion stage, followed by the first attempt of market entry, and, lastly, by the second attempt of market entry. The information problem between the entrepreneur and the investor implies that under short-term contracting good quality projects are stopped prematurely. The entrepreneur, therefore, seeks long-term financing. A long-term contract is more efficient and provides an "insurance" against the premature stop of the venture project.

(b) Distribution of the project's returns

How can investors recoup their funds? In case of debt financing, the bank obtains a fixed repayment and, thus, does not fully participate in the project's returns. Repayments in the first period are based on the firm's market position. The firm's market position, in turn, depends on success or failure of its innovation project as well as of that of the competitor. If a firm falls behind in the innovation game, the first-period profits will not suffice to cover the credit repayment. In this case, the bank denies follow-up financing for the firm. Repayments in the second period are all identical and amount to the lowest possible profits. The reason is that the bank has lost any threat potential, such that the entrepreneur is always induced to report low profits in the second period.

Since venture capital projects typically bear high failure risks, equity investors must fully participate in the project's upside potential in order to break even. Thus, the entrepreneur and the venture capitalist divide the project's gross profits. The optimal contract is a time-varying share contract. The share of profits accruing to the entrepreneur - and to the venture capitalist, respectively - is based on the scenario in place. In the first period, the share of profits that the entrepreneur obtains for incentive reasons is composed of various parts: The share of profits takes account of the static and the dynamic opportunities to misuse the funds, the learning rent, and the competition effect. Here, we derived the interesting result that the entrepreneur's first-period share of profits is higher under short-term than under long-term contracting. This implies that under long-term contracting the dynamic moral hazard problem is actually reduced. The effect stems from the repeated competition in the product market.

(c) Refinancing probabilities

In case of debt financing, refinancing probabilities between zero and one are an essential feature of the contract. The bank selects the refinancing probability according to the entrepreneur's reported profits. If reported profits are low, the bank will threaten to deny follow-up financing for the second period of competition. This helps to realign the incentives of the entrepreneur and induces truthful reporting of profits.

In case of equity financing, if venture capital is provided via a long-term contract, the refinancing probability will equal one. On the other hand, the refinancing probability equals zero if funds are provided via short-term contracts, and expected profits from a second attempt of market entry are too low.

(iv)   Agency costs of financial contracting

What are the consequences if the entrepreneur needs external equity or debt financing? In case of debt financing, the entrepreneur may not obtain refinancing for the second period, although the project is profitable in each period. A debt-financed firm, therefore, may have to stop production and to leave the market.

In case of venture capital financing, a long-term contract might not be offered when expected profits are too low to meet the incentive and participation constraints. If, in turn, only short-term contracts are offered, the entrepreneur may not obtain financing for the second attempt of market entry.

Next, we summarize our results on how these financial contracts will influence the firms' competition in the product market, their innovation behavior, and the market structure of the industry.

Impact of financial contracting on product market competition, innovation, and market structure

We now compare more in detail the impact of the financial contracts on *(i)* the firms' innovation behavior; *(ii)* the firms' pricing strategies; (iii) the market structure of the industry; *(iv)* the possibility for unfriendly strategic competition, i.e. predation.

(i)   Innovation activities

In our debt-financing model, R&D costs are quadratic and differ across periods. As for the innovation success probability in the first period, we derive that innovation activities will increase if the subsequent period of competition is taken into account. In the second period, the innovation activities are based on the market position acquired during the first period. The different profit opportunities imply that innovation success probability is higher for firms that start with a cost-advantage than for firms that start with a cost-disadvantage. This reflects the "success breeds success" hypothesis. In addition to that, we derive the interesting result that innovation activities are highest for firms in symmetric market positions. This holds as long as research costs are low and potential gains from innovation and cost-reduction are high.

In case of equity financing, the innovation technology is linear and identical for product and process innovation. Although this is a simple and very rudimentary formulation of the R&D technology, it still captures the essential features of innovation that we need in our product market competition game.

(ii) Pricing strategies

In case of debt-financing, we first recall that innovation causes marginal production costs to decrease from a high to a medium and then to a low level. Consequently, product prices will gradually be reduced.

If only one firm needs external debt financing, this firm's innovation activities will sharply decrease. Its product prices will, thus, be higher than those of its self-financed competitor. The self-financed rival dramatically increases its first-period R&D expenditures in order to obtain a cost-advantage and to acquire a large market share by charging low product prices. If this strategy is successful, the debt-financed firm has to leave the market because a bank does not refinance a firm with a cost-disadvantage. The self-financed firm, on the other hand, becomes monopolist and will charge the monopoly price.

Finally, if both firms need external debt financing, we show that the innovation success probabilities of both firms are reduced. Expected product prices in the first period are, thus, higher than under self-financing. Moreover, if one of the firms successfully innovates while the other does not, the unsuccessful firm obtains no refinancing and has to leave the market. The remaining firm will, in turn, charge the monopoly price in the second period.

We see that debt-financing softens price competition between firms because their innovation activities are reduced by the financial contract. In addition to this, with a certain probability consumers will have to pay the monopoly price in the second period.

In case of equity financing, the incumbent will continue to charge the monopoly price if the young firm fails to enter the market. If, by contrast, the young firm obtains venture capital financing and successfully innovates, expected product prices will fall below the original monopoly price.

(iii) Market structure

In case of debt financing, refinancing probabilities smaller than 1 may induce a credit-financed firm to exit. Thus, market structure may shift from a symmetric duopoly (if both firms need external debt-financing) or an asymmetric duopoly (if only one firm needs external debt-financing) to a monopoly in the product market.

In case of equity financing, if the young firm obtains venture capital and successfully enters the market, the intensity of competition will increase. Market structure shifts from monopoly to duopolistic price competition.

(iv) Predation

Lastly, we consider the issue of unfriendly strategic competition, i.e. predation. Here, a self-financed rival spends predation costs in order to reduce the success probability of the externally financed firm.

In case of debt financing, predation causes the leveraged firm's innovation probability to decline. Thus, it is very likely that this firm falls behind in the innovation and competition game. The bank, in turn, may deny refinancing. In this case, the firm has to leave the market, while the self-financed rival becomes monopolist. We conclude that financial debt contracts specifying a refinancing probability smaller than 1 make a firm vulnerable to predatory activities from part of the self-financed rival.

In case of equity financing and predation from part of the incumbent, the expected profits from the venture project will decline. As a consequence, the young firm may obtain only short-term financing or no financial contract at all. The incumbent, in turn, increases its likelihood of remaining monopolist. The predation strategy is especially successful if the young firm's new product is not very different from the existing one.

These findings confirm the well-known "deep-pocket" hypothesis from industrial organization. It states that firms with better access to financial resources, i.e. firms which dispose of more internal funds, will survive longer in competition.

Directions for future research

The present work started from the microeconomic level of an individual firm's financing decision. We then turned to the industry level where we analyzed how financial market imperfections influenced innovative activities and strategic competition of firms in the product market. A promising direction for future research would be to address now the macroeconomic level and to further investigate the macroeconomic consequences of these strategic input–output market interactions.

To date, only little work exists which investigates the consequences of capital market imperfections on economic growth: Bencivenga and Smith (1993) develop an endogenous growth model where all investment is financed via credit extension. An adverse selection problem in the credit market leads to credit rationing, which in turn impacts economic growth. Similarly, Ma and Smith (1996) consider a costly-state verification problem in the credit market. This imperfection of the credit market reduces the rate of economic growth. An excellent treatment of credit rationing and its impact on the business cycle and on economic growth is provided by Schubert (1999).

The same analysis is needed for equity financing. Here, a interesting first approach is given by King and Levine (1993). They construct an endogenous growth model in which financial systems, i.e. the stock markets, evaluate prospective entrepreneurs, and reveal the expected profits from innovation compared to the production of existing goods. The financial markets mobilize savings to finance the most productivity-enhancing activities. Better financial systems improve the probability of successful innovation and thereby accelerate economic growth. In reverse, King and Levine show that financial market imperfections reduce the rate of economic growth by reducing the rate of innovation.

Thus, in the long run, it would be most fascinating to develop a model that integrates all three levels: We would consider financial market imperfections due to asymmetric information on the input market side, market power of firms on the product market side and the respective linkages on economic growth, i.e. the steady state economy and the movement along the growth path (especially important for developing economies). The integration of imperfect financial and product markets will lead to a better and more detailed understanding of the growth process.

References

Admati, A., Pfleiderer, P., 1994. Robust financial contracting and the role of venture capitalists. Journal of Finance 49, 371—402.
Aghion, P., Bolton, P., 1992. An incomplete contracts approach to financial contracting. Review of Economic Studies 77, 338—401.
Akerlof, K., 1970. The market for "lemons": Quality uncertainty and the market mechanism. Quarterly Journal of Economics 84, 488—500.
Albach, H., Elston, J., 1995. Bank affiliations and firm capital investment in Germany. Ifo-Studien 41, 3—16.
Amit, R., Glosten, L., Muller, E., 1990. Entrepreneurial ability, venture investments, and risk sharing. Management Science 36, 1232—1245.
Anderlini, L., Felli, L., 1997. Costly Coasian contracts. Working paper, University of Cambridge.
Baltensperger, E., Devinney, T., 1985: Credit rationing theory: A survey and synthesis. Journal of Institutional and Theoretical Economics 141, 475—502.
Belleflamme, P., 2001. Oligopolistic competition, IT use for product differentiation and the productivity paradox. International Journal of Industrial Organization 19, 227—248.
Bencivenga, V., Smith, B. 1995. Transactions costs, technological choice, and endogenous growth. Journal of Economic Theory 67, 153—177.
Bencivenga, V., Smith, B. 1993. Some consequences of credit rationing in an endogenous growth model. Journal of Economic Dynamics and Control 17, 97—122.
Bergemann, D., Hege, U., 1998. Venture capital financing, moral hazard and learning. Journal of Banking and Finance 22, 703—735.
Berger, A., Udell, G., 1998. The economics of small business finance: The roles of private equity and debt markets in the financial growth cycle. Journal of Banking and Finance 22, 613—673.
Berglöf, E., 1994. A control theory of venture capital finance. Journal of Law, Economics, and Organization 10, 247—267.
Bester, H., 1994. The role of collateral in a model of debt renegotiation. Journal of Money, Credit and Banking 26, 72—86.
Bester, H., 1985a. Screening vs. rationing in credit markets with imperfect information. American Economic Review 75, 850—855.
Bester, H., 1985b. The level of investment in credit markets with imperfect information. Journal of Institutional and Theoretical Economics 141, 503—515.
Black, B., Gilson, R., 1998. Venture capital and the structure of capital markets: Banks versus stock markets. Journal of Financial Economics 47, 243—277.
Bolton, P., Scharfstein, D., 1996. Optimal debt structure and the number of creditors. Journal of Political Economy 104, 1—25.
Bolton, P., Scharfstein, D., 1990. A theory of predation based on agency problems in financial contracting. American Economic Review 80, 93—106.

Brander, J., Lewis, T.,1988. Bankruptcy costs and the theory of oligopoly. Canadian Journal of Economics 21, 221—243.

Brander, J., Lewis, T.,1986. Oligopoly and financial structure: The limited liability effect. American Economic Review 76, 957--970.

Bundesverband deutscher Kapitalbeteiligungsgesellschaften, 2000. Jahrbuch 1999. Berlin.

Bundesverband deutscher Kapitalbeteiligungsgesellschaften, 1999. Jahrbuch 1998. Berlin.

Casamatta, C., 1999. Financing and advising: Optimal financial contracts with venture capitalists. Working paper, University of Toulouse.

Chan, Y., Siegel, D., Thakor, A., 1990. Learning, corporate control and performance requirements in venture capital contracts. International Economic Review 31, 365—381.

Chang, C., 1990. The dynamic structure of optimal debt contracts. Journal of Economic Theory 52, 68—86.

Chevalier, J., 1995a. Capital structure and product market competition: Empirical evidence from the supermarket industry. American Economic Review 85, 415--435.

Chevalier, J., 1995b. Do LBO supermarkets charge more? An empirical analysis of the effect of LBOs on supermarket pricing. Journal of Finance 50, 1095--1112.

Chevalier, J., Scharfstein, D., 1996. Capital-market imperfections and countercyclical markups: Theory and evidence. American Economic Review 86, 703--725.

Clemenz, G., 1986. Credit Markets with Asymmetric Information. Lecture Notes in Economics and Mathematical Systems. Berlin, Heidelberg, New York, Springer.

Coase, R., 1937. The nature of the firm. Economica 4, 386—405. Reprinted in Stigler, G., Boulding, K., 1952. Readings in price theory. Homewood, Irwin (for American Economic Association).

Cornelli, F., Yosha, O., 1997. Staged financing and the role of convertible debt. CEPR Discussion Paper 1735.

D'Aspremont, C., Gabczewicz, J., Thisse, J., 1979. On Hotelling's stability in competition. Econometrica 17, 1145—1151.

Dasgupta, S., Titman, S., 1998. Pricing strategy and financial policy. Review of Financial Studies 11, 705—737.

Dasgupta, S., Titman, S., 1996. Pricing strategy and financial policy. Working Paper 5498, National Bureau of Economic Research.

De Meza, D., Webb, D., 1992. Efficient credit rationing. European Economic Review 36, 1277-1290.

De Meza, D., Webb, D., 1987. Too much investment: A problem of asymmetric information. Quarterly Journal of Economics 102, 281—292.

Deutsche Bundesbank, 2001. Monatsbericht März 2001.

Diamond, D., 1991. Debt maturity structure and liquidity risk. Quarterly Journal of Economics 106, 709–737.

Diamond, D., 1989. Reputation acquisition in debt markets. Journal of Political Economy 97, 828--862.

Diamond, D., 1984. Financial intermediation and delegated monitoring. Review of Economic Studies 51, 393—414.

Elsas, R., Krahnen, J., 1998. Is relationship lending special? Evidence from credit-file data in Germany. Journal of Banking and Finance 22, 1283–1316.

Fama, E., Miller, M., 1972. The theory of finance. Holt, Rinehart, and Winston.

Faure-Grimaud, A., 2000. Product market competition and optimal debt contracts: The limited liability effect revisited. European Economic Review 44, 1823—1840.

Fazzari, S., Hubbard, G., Petersen, B., 1988. Financing constraints and corporate investment. Brookings Papers on Economic Activity, 141—195.

Fenn, G., Liang, N., 1998. New resources and new ideas: Private equity for small business. Journal of Banking and Finance 22, 1077—1084.

Fenn, G., Liang, N., Prowse, S., 1997. The private equity market: an overview. Financial Markets Institutions and Instruments 6, 1–105

Fluck, Z., 1998. Optimal financial contracting: Debt versus outside equity. Review of Financial Studies 11, 383—418.
Freixas, X., Rochet, J., 1998. Microeconomics of banking. Cambridge, MIT Press, 2nd edition.
Fudenberg, D., Tirole, J., 1986. A signal jamming theory of predation. Rand Journal of Economics 17, 366—376.
Fudenberg, D., Tirole, J., 1990. Moral hazard and renegotiation in agency contracts. Econometrica 58, 1279—1319.
Fudenberg, D., Tirole, J., 1991. Game Theory. Cambridge, MIT Press.
Gale, D., Hellwig, M., 1985. Incentive-compatible debt contracts: The one-period problem. Review of Economic Studies 52, 647--663.
Gertner, R., Gibbons, R., Scharfstein, D., 1988. Simultaneous signaling to the capital and product markets. Rand Journal of Economics 19, 173—190.
Glazer, J., 1994. The strategic effects of long-term debt. Journal of Economic Theory 62, 428--443.
Glazer, J., Israel, R., 1990. Managerial incentives and financial signaling in product market competition. International Journal of Industrial Organization 8, 271—280.
Gompers, P., 1998. Venture capital growing pains: should the market diet? Journal of Banking and Finance 22, 1089—1104.
Gompers, P., 1995. Optimal investment, monitoring, and the staging of venture capital. Journal of Finance 50, 1461—1489.
Gompers, P., Lerner, J., 1999a. An analysis of compensation in the U.S. venture capital partnerships. Journal of Financial Economics 51, 3-44.
Gompers, P., Lerner, J., 1999b. What drives venture capital fundraising? NBER Working Paper 6906.
Gompers, P., Lerner, J., 1999c. The venture capital cycle. MIT Press
Gompers, P., Lerner, J., 1996. The use of covenants: an empirical analysis of venture partnership agreements. Journal of Law and Economics 39, 463—498.
Gorman, M., Sahlman, W., 1989. What do venture capitalists do? Journal of Business Venturing 4, 213—248.
Gottardi, P., 1995. An analysis of the conditions for the validity of the Modigliani-Miller theorem with incomplete markets. Journal of Economic Theory 5, 189—207.
Grossman, G., Hart, O., 1986. The costs and benefits of ownership: A theory of vertical and lateral integration. Journal of Political Economy 94, 691—719.
Grossman, G., Hart, O., 1983. An analysis of the principal-agent problem. Econometrica 51, 7—45.
Hansen, E., 1991. Venture capital finance with temporary asymmetric learning. Financial Market Group Discussion Paper 112, London School of Economics.
Harhoff, D., Körting, T., 1998. Lending relationships in Germany: Empirical results from survey data. Journal of Banking and Finance 22, 1317—1353.
Harris, M., Raviv, A., 1991. The theory of capital structure. Journal of Finance 46, 297—355.
Harris, M., Raviv, A., 1990. Capital structure and the informational role of debt. Journal of Finance 45, 321—349.
Harris, M., Raviv, A., 1989. Corporate control contests and capital structure. Journal of Financial Economics 20, 55—86.
Hart, O., 1995. Firms, Contracts, and Financial Structure. Oxford, Oxford University Press.
Hellmann, T., 1998. The allocation of control rights in venture capital contracts. Rand Journal of Economics 29, 57—76.
Hellmann, T., Puri, M., 1999. The interaction between product market and financial strategy: The role of venture capital. Working paper, Stanford University.
Hellwig, M., 1998. On the economics of corporate finance and corporate control. Working paper University of Mannheim.
Holmström, B., 1982. Moral hazard in teams. Bell Journal of Economics 13, 324—340.
Holmström, B., 1979. Moral hazard and observability. Bell Journal of Economics 10, 74—91.

Hubbard, G., 1998. Capital-market imperfections and investment. Journal of Economic Literature 36, 193-225.

Hubert, F., 1999. Optimale Finanzkontrakte, Investitionspolitik und Wettbewerbskraft. Heidelberg, Physica.

Jaffee, D., Stiglitz, J., 1990. Credit rationing. In: Friedman, M., Hahn, F., (Eds): Handbook of monetary economics. Amsterdam, Elsevier, chapter 16.

Jagannathan, R., Srinivasan, S., 2000. Does product market competition reduce agency costs? Working Paper 7480, National Bureau of Economic Research.

Jensen, M., 1986. Agency costs of free cash flow, corporate finance, and takeovers. American Economic Review Papers and Proceedings 76, 375—394.

Jensen, M., Meckling, W., 1976. The theory of the firm: Managerial behavior, agency costs, and ownership structure. Journal of Financial Economics 3, 305—360.

Jewitt, I., 1988. Justifying the first-order approach to principal-agent problems. Econometrica 56, 1177—1190.

Kaplan, S., Strömberg, P., 2000. Financial contracting theory meets the real world: An empirical analysis of venture capital contracts. Working Paper 7660, National Bureau of Economic Research.

King, R., Levine, R., 1993. Finance, entrepreneurship, and growth. Theory and evidence. Journal of Monetary Economics 32, 523—542.

Klein, B., Crawford, R., Alchian, A., 1978. Vertical integration, appropriable rents, and the competitive contracting process. Journal of Law and Economics 21, 297–326.

Klemperer, P., 1997. Markets with consumer switching costs. Quarterly Journal of Economics 102, 375-394.

Klemperer, P., 1995. Competition when consumers have switching costs: An overview with applications to industrial organization, macroeconomics, and international trade. Review of Economic Studies 62, 515--539.

Kovenock, D., Phillips, G., 1997. Capital structure and product market behavior: An examination of plant exit and investment decisions. Review of Financial Studies 10, 767—803.

Kovenock, D., Phillips, G., 1995. Capital structure and product market rivalry: How do we reconcile theory and evidence. American Economic Review 85, 403—408.

Lehmann, E., Neuberger, D., 1998. SME loan pricing and lending relationships in Germany: A new look. Working Paper, University of Rostock.

Leland, H., Pyle, D., 1977. Informational asymmetries, financial structure, and financial intermediation. Journal of Finance 32, 371--387.

Lerner, J., 1998. "Angel" financing and public policy: An overview. Journal of Banking and Finance 22, 773—783.

Lerner, J., 1998. Comment on Bergemann and Hege. Journal of Banking and Finance 22, 736—740.

Ma, C., Smith, B., 1996. Credit market imperfections and economic development: theory and evidence. Journal of Development Economics 48, 351—387.

Maksimovic, V., 1990. Product market imperfections and loan commitments. Journal of Finance 45, 1641--1653.

Marx, L., 1998. Efficient venture capital financing combining debt and equity. Review of Economic Design 3, 371—387.

Mas Colell, A., Whinston, M., Green, J., 1995. Microeconomic Theory. Oxford, Oxford University Press.

Maurer, B., 1999. Innovation and investment under financial constraints and product market competition. International Journal of Industrial Organization 17, 455--476.

Melumad, N., Mookherjee, D., Reichelstein, S., 1995. Hierarchical decentralization of incentive contracts. Rand Journal of Economics 26, 654—672.

Milde, H., Riley, J., 1988. Signaling in credit markets. Quarterly Journal of economics 103, 101—130.

Milgrom, P., Roberts, P., 1982. Limit pricing and entry under incomplete information: an equilibrium analysis. Econometrica 50, 443—460.

Mirrlees, J., 1997. Information and incentives: The economics of carrots and sticks. Economic Journal 107, 105—131.
Mirrlees, J., 1976. The optimal structure of incentives and authority within an organization. Bell Journal of Economics 7, 105—131.
Mishkin, F., 1995. Symposium on the monetary transmission mechanism. Journal of Economic Perspectives 9, 3--10.
Modigliani, F., 1988. MM – past, present, future. Journal of Economic Perspectives 2, 149—158.
Modigliani, F., Miller, M., 1963. Corporate income taxes and the cost of capital: A correction. American Economic Review 53,
Modigliani, F., Miller, M., 1958. The cost of capital, corporation finance, and the theory of investment. American Economic Review 48, 261—297.
Mookherjee, D., 1999. The value of delegation. Oberwesel, International Summer School on Interactive Economic Decisions lecture notes.
Myers, S., 2000. Outside equity financing. Journal of Finance 55, 1005--1037.
Myers, S., 1984. Presidential address: The capital structure puzzle. Journal of Finance 39, 575—592.
Myers, S., Majluf, N., 1984. Corporate financing and investment decisions when firms have information that investors do not have. Journal of Financial Economics 13, 187--221.
Nachman, D., Noe, T. 1994. Optimal design of securities under asymmetric information. Review of Financial Studies 7, 1--44.
Neff, C., 1999. How deep is the pocket? Agency problems in financial contracting and their impact on innovation and strategic competition. Working Paper 129, University of Tübingen.
Neff, C., 1997. Finanzstruktur und strategischer Wettbewerb auf Gütermärkten. Working Paper 89, University of Tübingen.
Opler, T., Titman, S., 1994. Financial distress and corporate performance. Journal of Finance 49, 1015--1040
Organisation for Economic Co-operation and Development, 1996. Venture capital and innovation. Paris, Organisation for Economic Co-operation and Development.
Pfirrmann, O., Wupperfeld, U., Lerner, J., 1997. Venture capital and new technology based firms: an US-German comparison. Heidelberg, Physica.
Phillips, G., 1995. Increased debt and industry product markets: An empirical analysis. Journal of Financial Economics 37, 189—238.
Poitevin, M., 1989. Financial signalling and the "deep-pocket" argument. Rand Journal of Economics 20, 26-40.
Pratt, J., Zeckhauser, R., 1985. Principals and agents: An overview. In: Pratt, J., Zeckhauser, R.: Principals and Agents: The Structure of Business. Boston.
Ramakrishnan, S., Thakor, A., 1984. Information reliability and a theory of financial intermediation. Review of Economic Studies 51, 415—432.
Ramser, H., Stadler, M., 1995. Kreditmärkte und Innovationsaktivität. Ifo-Studien 41, 187—207.
Rasmusen, E., 1994. Games and Information. Cambridge, 2nd Edition.
Ravid, A., 1988. On interactions of production and financial decisions. Financial Management, 87—99.
Reichelstein, S., 1999. Incentives and organization design. Oberwesel, International Summer School 1999 lecture notes.
Reid, G., 1996. Fast growing small entrepreneurial firms and their venture capital backers: An applied principal-agent analysis. Small Business Economics 8, 235—248.
Repullo, R., Suarez, J., 2000. Entrepreneurial moral hazard and bank monitoring: A model of the credit channel. European Economic Review 44, 1931--1950.
Repullo, R., Suarez, J., 1998. Venture capital finance: A security design approach. Working paper, University of Madrid.
Ross, S., 1977. The determination of financial structure: The incentive-signalling approach. Bell Journal of Economics 8, 23—40.
Rothschild, M., Stiglitz, J., 1970. Increasing risk: A definition. Journal of Economic Theory 2, 225—243.

Ruckes, M., 1998. Competing banks, credit standards and corporate conservatism. Working Paper, University of Mannheim.

Sahlman, W., 1990. The structure and governance of venture capital organizations. Journal of Financial Economics 27, 473—521.

Salop, S., 1979. Monopolistic competition with outside goods. Bell Journal of Economics 10, 141—156.

Schefzcyck, M., 2000. Erfolgsstrategien deutscher Venture Capital-Gesellschaften – Analyse der Investitionsaktivitäten und des Beteiligungsmanagement von Venture Capital-Gesellschaften. Stuttgart, Schäffer-Poeschel, 2nd ed.

Schmidt, K., 2000. Anreizprobleme bei der Finanzierung mit Wagniskapital. In: Franz, W., Hesse, H., Ramser, HJ., Stadler, M., (Eds.). Ökonomische Analyse von Verträgen. Wirtschaftswissenschaftliches Seminar Ottobeuren 29. Tübingen, Mohr Siebeck, 248—284.

Schnitzer, M., Wambach, A., 1998. Inside versus outside financing and product market competition. Working paper, University of München.

Schubert, V., 1999. Asymmetrische Information und Finanzierungsstruktur: Rationierung auf dem Kapitalmarkt: empirische Befunde, informationsökonomische Mikrofundierung und makroökonomische Modellierung. Berlin, Duncker & Humblot.

Schweizer, U., 1999. Vertragstheorie. Tübingen, Mohr Siebeck.

Showalter, D., 1995. Oligopoly and financial structure: Comment. American Economic Review 85, 647–653.

Shy, O., 1995. Industrial Organization – Theory and Applications. Cambridge, MIT Press.

Sidler, S., 1997. Risikokapitalfinanzierung von Jungunternehmen. Berlin Stuttgart Wien, Haupt, 2nd ed.

Sinclair-Desgagne, B., 1994. The first-order approach to multi-signal principal agent problems: Notes and comments. Econometrica 62, 459—465.

Smith, A. 1937. The Wealth of Nations. New York, Random House (first published 1776).

Snyder, C., 1996. Negotiation and renegotiation of optimal financial contracts under the threat of predation. Journal of Industrial Economics 44, 325—343.

Stadler, M., 1997. Two-period financial contracts and product market competition. Ifo-Studien 43, 367–381.

Stiglitz, J., 1988. Why financial structure matters. Journal of Economic Perspectives, 121-126.

Stiglitz, J., Weiss, A., 1986. Credit rationing and collateral. In: Edwards, J., Franks, J., Mayer, C., Schaefer, S., (Eds.). Recent Developments in Corporate Finance. New York, Cambridge University Press, 101--135.

Stiglitz, J., Weiss, A., 1981. Credit rationing in markets with imperfect information. American Economic Review 71, 393--410.

Sutton, J., 1998. Technology and Market Structure. Cambridge, MIT Press.

Telser, L., 1966. Cutthroat competition and the long purse. Journal of Law and Economics 9, 259—277.

Thakor, A., 1998. Comment on Trester. Journal of Banking and Finance 22, 700—701.

Tirole, J., 1999. Incomplete contracts: where do we stand? Econometrica 67, 741—781.

Tirole, J., 1988. The Theory of Industrial Organization. Cambridge, MIT Press.

Townsend, R., 1979. Optimal contracts and competitive markets with costly state verifications. Journal of Economic Theory 21, 265--293.

Trester, J., 1998. Venture capital contracting under asymmetric information. Journal of Banking and Finance 22, 675—699.

Webb, D., 1992. Two-period financial contracts with private information and costly state verification. Quarterly Journal of Economics 107,1113—1123.

Webb, D., 1991. Long-term financial contracts can mitigate the adverse selection problem in project financing. International Economic Review 32, 305--320.

Wette, H., 1983. Collateral in credit rationing in markets with imperfect information: A note. American Economic Review 73, 442—445.

Williamson, O., 1979. Predatory pricing. A strategic and welfare analysis. Economic Analysis and Antitrust Law, 195—239.

Williamson, O., 1985. Assessing contracts. Journal of Law, Economics and Organization 1, 177—208.
Williamson, O., 1987. Costly monitoring, loan contracts, and equilibrium credit rationing. Quarterly Journal of Economics 102, 135--145.
Williamson, O., 1988. Corporate finance and corporate governance. Journal of Finance 43, 567—591.
Winker, P., 1996. Rationierung auf dem Markt für Unternehmenskredite in der BRD. Tübingen, Mohr Siebeck.
Zechner, J., 1996. Financial market-product market interactions in industry equilibrium: Implications for information acquisition decisions. European Economic Review 40, 883–896.
Zingales, L., 1998. Survival of the fittest or the fattest? Exit and financing in the trucking industry. Journal of Finance 53, 905—938.

Figures

2.1	Effect of an increase in debt under demand-uncertainty	11
2.2	Effect of an increase in debt under cost-uncertainty	12
2.3	Product market equilibria if firms compete in prices and take up strategic debt	13
2.4	Outward shift of the capacity reaction function due to an increase of debt	18
2.5	Output market equilibrium with and without firms taking up debt	21
2.6	The effect of long-term debt on first-period prices	25
2.7	The Hotelling-line	27
2.8	Debt-induces predation from part of the incumbent firm	40
3.1	The standard debt contract	53
3.2	Moral hazard and renegotiation of the debt contract	55
3.3	Price reaction curves for firms i,j with self-financing and external debt-financing	67
3.4	Research costs and innovation success probability	78
3.5	R&D reaction functions in the basic game	81
3.6	Innovation success probabilities and payoffs in the two-period case	83
3.7	R&D reaction functions in the two-period game	87
3.8	R&D reaction functions if firm i is externally financed via a long-term loan contract and $\beta^S=1$ (Case 1)	93
3.9	R&D reaction functions if firm i is externally financed via a long-term loan contract and $\beta^S<1$ (Case 2)	93
3.10	R&D reaction functions if both firms are loan-financed and refinancing probabilities equal $\beta^S=\gamma^S=1$ (Case 3)	98
3.11	R&D reaction functions if both firms are loan-financed and refinancing probabilities equal $\beta^S<1, \gamma^S<1$ (Case 4)	99
4.1	The financial growth cycle	108
4.2	Exit channels for venture capital investors	112
4.3	Project value under venture capital financing and double moral hazard	134
4.4	Time-line of the basic entry game	157
4.5	The two-period game of innovation and price competition	159
4.6	First-period payoffs when both firms spend on R&D activities	180
4.7	Payoffs in the two-period game when the incumbent engages in predation	185

Tables

2.1	Payoffs according to the firms' respective debt levels	20
3.1	Prices, market shares, and profits after the first period of innovation.	80
3.2	Second-period gross profits if firm i starts with a cost-disadvantage..	84
3.3	Second period gross profits if firm i starts with a cost-advantage	84
3.4	Optimal contract in case of a low innovation probability	91
3.5	Optimal contract in case of a high innovation probability	92
4.1	Multi-stage selection process before venture capital contracts are signed	111
4.2	Second-period payoffs after symmetric costs in $t=1$	180
4.3	Second-period payoffs following a cost-disadvantage of the young firm in $t=1$	181
4.4	Second-period payoffs after the young firm failed to enter in $t=1$ and the incumbent produced with high marginal costs	181
4.5	Second-period payoffs after the young firm failed to enter in $t=1$ and the incumbent produced with medium marginal costs	182

List of symbols

LATIN ALPHABET

A	Assets pledged as collateral
B	Benefit from a bankruptcy of the rival firm
C	Collateral
D	Debt level or loan size
E	Expectation operator
F	Cumulated distribution function for discrete variable
I	Investment level
J	Variable indicating whether costly monitoring takes place
K	Capacity (in chapter 2)
K^{pred}	Predation costs (in chapters 3 and 4)
L	Liquidation value
M	Indicator for who is in control of the venture project
N	Number of shareholders
P	Point in figure
Q	Market demand
R	Repayment obligation
S	Fraction of shares or profits
T	Transportation costs or consumer's disutility
U	Utility function
V	Firm value accruing to entrepreneur
W	Firm value accruing to investor
Y	Total firm value combining debt and equity
a	Number of shares
b	Bankruptcy costs
c	Marginal production cost
c_K	Marginal cost of capacity
e	Effort
f	Density function for discrete variable
g	Parameter of research cost function
i	Index for incumbent i
j	Index for firm j
k	Inverse function
l	Parameter below which monitoring takes place in Chang's model
m	Monitoring costs
n	Number of firms or innovation projects
p	Price
p^r	Reservation price
r	Interest rate
s	Consumers' reservation value in the Hotelling model

List of symbols

t	Time index
u	Utility function
v	Disutility of effort
w	Fraction of the firm's asset value in the hands of the bank after liquidation
x	Consumer's location in the Hotelling-model
y	Index for young firm
z	Random variable representing state of nature

GREEK ALPHABET

Γ	Contract
Δ	Difference in cost
Π	Profits
Φ	Cumulative distribution function for continuous variable
Ψ	Threshold parameter in Casamatta's model
Ω	Abbreviated function in Casamatta's model
α	Beliefs about the project's quality
β	Refinancing probability of firm i
γ	Refinancing probability for rival firm
δ	Discount factor
ε	Potential second-period profits if firm obtains financing only in the first period
ζ	Reaction function parameter
η	Bankruptcy vs. renegotiation decision in Bester's model
θ	Success probability of the innovation project
ϑ	Reaction function parameter
ι	Reaction function parameter
κ	Number of shares
λ	Innovation success probability of rival firm in chapter 3
μ	Second-period innovation success probability in chapter 4
ξ	Venture capitalist's fraction of second-period investment in Admati and Pfleiderer's model
ρ	Parameter of the R&D cost function
σ	Market share
τ	Truth
φ	Density function of continuous variable
χ	Proportionality factor of the reservation price
ψ	Inverse function
ω	Initial wealth of the entrepreneur

Mathematical Symbols

$\mathcal{R}$	Set of real numbers
$\mathcal{N}$	Set of positive integers
$\equiv$	Definition
$\approx$	Is proportional to
$\Leftrightarrow$	If and only if
$\in$	Is an element of
$\{\ \vert\ \}$	Given that
$\wedge$	Indicates the agent's report of a privately observable variable

Druck und Bindung: Strauss Offsetdruck GmbH

Lecture Notes in Economics and Mathematical Systems

For information about Vols. 1–429
please contact your bookseller or Springer-Verlag

Vol. 430: J. R. Daduna, I. Branco, J. M. Pinto Paixão (Eds.), Computer-Aided Transit Scheduling. XIV, 374 pages. 1995.

Vol. 431: A. Aulin, Causal and Stochastic Elements in Business Cycles. XI, 116 pages. 1996.

Vol. 432: M. Tamiz (Ed.), Multi-Objective Programming and Goal Programming. VI, 359 pages. 1996.

Vol. 433: J. Menon, Exchange Rates and Prices. XIV, 313 pages. 1996.

Vol. 434: M. W. J. Blok, Dynamic Models of the Firm. VII, 193 pages. 1996.

Vol. 435: L. Chen, Interest Rate Dynamics, Derivatives Pricing, and Risk Management. XII, 149 pages. 1996.

Vol. 436: M. Klemisch-Ahlert, Bargaining in Economic and Ethical Environments. IX, 155 pages. 1996.

Vol. 437: C. Jordan, Batching and Scheduling. IX, 178 pages. 1996.

Vol. 438: A. Villar, General Equilibrium with Increasing Returns. XIII, 164 pages. 1996.

Vol. 439: M. Zenner, Learning to Become Rational. VII, 201 pages. 1996.

Vol. 440: W. Ryll, Litigation and Settlement in a Game with Incomplete Information. VIII, 174 pages. 1996.

Vol. 441: H. Dawid, Adaptive Learning by Genetic Algorithms. IX, 166 pages. 1996.

Vol. 442: L. Corchón, Theories of Imperfectly Competitive Markets. XIII, 163 pages. 1996.

Vol. 443: G. Lang, On Overlapping Generations Models with Productive Capital. X, 98 pages. 1996.

Vol. 444: S. Jørgensen, G. Zaccour (Eds.), Dynamic Competitive Analysis in Marketing. X, 285 pages. 1996.

Vol. 445: A. H. Christer, S. Osaki, L. C. Thomas (Eds.), Stochastic Modelling in Innovative Manufactoring. X, 361 pages. 1997.

Vol. 446: G. Dhaene, Encompassing. X, 160 pages. 1997.

Vol. 447: A. Artale, Rings in Auctions. X, 172 pages. 1997.

Vol. 448: G. Fandel, T. Gal (Eds.), Multiple Criteria Decision Making. XII, 678 pages. 1997.

Vol. 449: F. Fang, M. Sanglier (Eds.), Complexity and Self-Organization in Social and Economic Systems. IX, 317 pages, 1997.

Vol. 450: P. M. Pardalos, D. W. Hearn, W. W. Hager, (Eds.), Network Optimization. VIII, 485 pages. 1997.

Vol. 451: M. Salge, Rational Bubbles. Theoretical Basis, Economic Relevance, and Empirical Evidence with a Special Emphasis on the German Stock Market. IX, 265 pages. 1997.

Vol. 452: P. Gritzmann, R. Horst, E. Sachs, R. Tichatschke (Eds.), Recent Advances in Optimization. VIII, 379 pages. 1997.

Vol. 453: A. S. Tangian, J. Gruber (Eds.), Constructing Scalar-Valued Objective Functions. VIII, 298 pages. 1997.

Vol. 454: H.-M. Krolzig, Markov-Switching Vector Autoregressions. XIV, 358 pages. 1997.

Vol. 455: R. Caballero, F. Ruiz, R. E. Steuer (Eds.), Advances in Multiple Objective and Goal Programming. VIII, 391 pages. 1997.

Vol. 456: R. Conte, R. Hegselmann, P. Terna (Eds.), Simulating Social Phenomena. VIII, 536 pages. 1997.

Vol. 457: C. Hsu, Volume and the Nonlinear Dynamics of Stock Returns. VIII, 133 pages. 1998.

Vol. 458: K. Marti, P. Kall (Eds.), Stochastic Programming Methods and Technical Applications. X, 437 pages. 1998.

Vol. 459: H. K. Ryu, D. J. Slottje, Measuring Trends in U.S. Income Inequality. XI, 195 pages. 1998.

Vol. 460: B. Fleischmann, J. A. E. E. van Nunen, M. G. Speranza, P. Stähly, Advances in Distribution Logistic. XI, 535 pages. 1998.

Vol. 461: U. Schmidt, Axiomatic Utility Theory under Risk. XV, 201 pages. 1998.

Vol. 462: L. von Auer, Dynamic Preferences, Choice Mechanisms, and Welfare. XII, 226 pages. 1998.

Vol. 463: G. Abraham-Frois (Ed.), Non-Linear Dynamics and Endogenous Cycles. VI, 204 pages. 1998.

Vol. 464: A. Aulin, The Impact of Science on Economic Growth and its Cycles. IX, 204 pages. 1998.

Vol. 465: T. J. Stewart, R. C. van den Honert (Eds.), Trends in Multicriteria Decision Making. X, 448 pages. 1998.

Vol. 466: A. Sadrieh, The Alternating Double Auction Market. VII, 350 pages. 1998.

Vol. 467: H. Hennig-Schmidt, Bargaining in a Video Experiment. Determinants of Boundedly Rational Behavior. XII, 221 pages. 1999.

Vol. 468: A. Ziegler, A Game Theory Analysis of Options. XIV, 145 pages. 1999.

Vol. 469: M. P. Vogel, Environmental Kuznets Curves. XIII, 197 pages. 1999.

Vol. 470: M. Ammann, Pricing Derivative Credit Risk. XII, 228 pages. 1999.

Vol. 471: N. H. M. Wilson (Ed.), Computer-Aided Transit Scheduling. XI, 444 pages. 1999.

Vol. 472: J.-R. Tyran, Money Illusion and Strategic Complementarity as Causes of Monetary Non-Neutrality. X, 228 pages. 1999.

Vol. 473: S. Helber, Performance Analysis of Flow Lines with Non-Linear Flow of Material. IX, 280 pages. 1999.

Vol. 474: U. Schwalbe, The Core of Economies with Asymmetric Information. IX, 141 pages. 1999.

Vol. 475: L. Kaas, Dynamic Macroeconomics with Imperfect Competition. XI, 155 pages. 1999.

Vol. 476: R. Demel, Fiscal Policy, Public Debt and the Term Structure of Interest Rates. X, 279 pages. 1999.

Vol. 477: M. Théra, R. Tichatschke (Eds.), Ill-posed Variational Problems and Regularization Techniques. VIII, 274 pages. 1999.

Vol. 478: S. Hartmann, Project Scheduling under Limited Resources. XII, 221 pages. 1999.

Vol. 479: L. v. Thadden, Money, Inflation, and Capital Formation. IX, 192 pages. 1999.

Vol. 480: M. Grazia Speranza, P. Stähly (Eds.), New Trends in Distribution Logistics. X, 336 pages. 1999.

Vol. 481: V. H. Nguyen, J. J. Strodiot, P. Tossings (Eds.). Optimation. IX, 498 pages. 2000.

Vol. 482: W. B. Zhang, A Theory of International Trade. XI, 192 pages. 2000.

Vol. 483: M. Königstein, Equity, Efficiency and Evolutionary Stability in Bargaining Games with Joint Production. XII, 197 pages. 2000.

Vol. 484: D. D. Gatti, M. Gallegati, A. Kirman, Interaction and Market Structure. VI, 298 pages. 2000.

Vol. 485: A. Garnaev, Search Games and Other Applications of Game Theory. VIII, 145 pages. 2000.

Vol. 486: M. Neugart, Nonlinear Labor Market Dynamics. X, 175 pages. 2000.

Vol. 487: Y. Y. Haimes, R. E. Steuer (Eds.), Research and Practice in Multiple Criteria Decision Making. XVII, 553 pages. 2000.

Vol. 488: B. Schmolck, Ommitted Variable Tests and Dynamic Specification. X, 144 pages. 2000.

Vol. 489: T. Steger, Transitional Dynamics and Economic Growth in Developing Countries. VIII, 151 pages. 2000.

Vol. 490: S. Minner, Strategic Safety Stocks in Supply Chains. XI, 214 pages. 2000.

Vol. 491: M. Ehrgott, Multicriteria Optimization. VIII, 242 pages. 2000.

Vol. 492: T. Phan Huy, Constraint Propagation in Flexible Manufacturing. IX, 258 pages. 2000.

Vol. 493: J. Zhu, Modular Pricing of Options. X, 170 pages. 2000.

Vol. 494: D. Franzen, Design of Master Agreements for OTC Derivatives. VIII, 175 pages. 2001.

Vol. 495: I Konnov, Combined Relaxation Methods for Variational Inequalities. XI, 181 pages. 2001.

Vol. 496: P. Weiß, Unemployment in Open Economies. XII, 226 pages. 2001.

Vol. 497: J. Inkmann, Conditional Moment Estimation of Nonlinear Equation Systems. VIII, 214 pages. 2001.

Vol. 498: M. Reutter, A Macroeconomic Model of West German Unemployment. X, 125 pages. 2001.

Vol. 499: A. Casajus, Focal Points in Framed Games. XI, 131 pages. 2001.

Vol. 500: F. Nardini, Technical Progress and Economic Growth. XVII, 191 pages. 2001.

Vol. 501: M. Fleischmann, Quantitative Models for Reverse Logistics. XI, 181 pages. 2001.

Vol. 502: N. Hadjisavvas, J. E. Martínez-Legaz, J.-P. Penot (Eds.), Generalized Convexity and Generalized Monotonicity. IX, 410 pages. 2001.

Vol. 503: A. Kirman, J.-B. Zimmermann (Eds.), Economics with Heterogenous Interacting Agents. VII, 343 pages. 2001.

Vol. 504: P.-Y. Moix (Ed.),The Measurement of Market Risk. XI, 272 pages. 2001.

Vol. 505: S. Voß, J. R. Daduna (Eds.), Computer-Aided Scheduling of Public Transport. XI, 466 pages. 2001.

Vol. 506: B. P. Kellerhals, Financial Pricing Models in Continuous Time and Kalman Filtering. XIV, 247 pages. 2001.

Vol. 507: M. Koksalan, S. Zionts, Multiple Criteria Decision Making in the New Millenium. XII, 481 pages. 2001.

Vol. 508: K. Neumann, C. Schwindt, J. Zimmermann, Project Scheduling with Time Windows and Scarce Resources. XI, 335 pages. 2002.

Vol. 509: D. Hornung, Investment, R&D, and Long-Run Growth. XVI, 194 pages. 2002.

Vol. 510: A. S. Tangian, Constructing and Applying Objective Functions. XII, 582 pages. 2002.

Vol. 511: M. Külpmann, Stock Market Overreaction and Fundamental Valuation. IX, 198 pages. 2002.

Vol. 512: W.-B. Zhang, An Economic Theory of Cities.XI, 220 pages. 2002.

Vol. 513: K. Marti, Stochastic Optimization Techniques. VIII, 364 pages. 2002.

Vol. 514: S. Wang, Y. Xia, Portfolio and Asset Pricing. XII, 200 pages. 2002.

Vol. 515: G. Heisig, Planning Stability in Material Requirements Planning System. XII, 264 pages. 2002.

Vol. 516: B. Schmid, Pricing Credit Linked Financial Instruments. X, 246 pages. 2002.

Vol. 517: H. I. Meinhardt, Cooperative Decision Making in Common Pool Situations. VIII, 205 pages. 2002.

Vol. 518: S. Napel, Bilateral Bargaining. VIII, 188 pages. 2002.

Vol. 519: A. Klose, G. Speranza, L. N. Van Wassenhove (Eds.), Quantitative Approaches to Distribution Logistics and Supply Chain Management. XIII, 421 pages. 2002.

Vol. 520: B. Glaser, Efficiency versus Sustainability in Dynamic Decision Making. IX, 252 pages. 2002.

Vol. 521: R. Cowan, N. Jonard (Eds.), Heterogenous Agents, Interactions and Economic Performance. XIV, 339 pages. 2003.

Vol. 522: C. Neff, Corporate Finance, Innovation, and Strategic Competition. IX, 218 pages. 2003.